J.T. Fraser
Time as Conflict

Aphrodite and Pan Playing Five Stones. Inside of a mirror cover from the Collection of Bronzes, the British Museum. Courtesy of the Trustees.

THIS DELIGHTFUL ENGRAVING is on a Greek Bronze mirror. It is said to come from Corinth, and is dated 350–300 B.C.

Aphrodite and Pan are playing five stones, a game of chance, which was originally a magical means to divine the future. A primitive level of the world is also controlled by chance, or laws of probability. In that world ideas of now versus then are only very loosely applicable, if at all. We shall call such worlds prototemporal.

The beauty and the beast are seated on a stone slab, a kind of bulk matter that makes up the astronomical universe of galaxies – albeit in many forms and at many temperatures. The time of that universe does not have a preferred direction and is without a definable present. We shall call such worlds eotemporal.

The goose in front of the bench is a traditional attribute of Aphrodite, as are doves, swans, scallop shells, and dolphins. It is the necessary internal coherence of life that defines a "now" in the eotemporal world of inanimate matter. Biotemporality is the name we give to the time of mindless life, the time mode manifest by organic evolution.

Pan is a woodland god of pastures and flocks who lives in the romantic paradise of Arcadia. He plays his reed pipes and aids hunters, but is also the cause of sudden, unreasonable feelings of terror. As a creature between two stable forms of nature (man and animal) he represents for us a family of metastable conditions. Such conditions separate the stable integrative levels of nature, each with its distinct temporality, causation, and language.

The central figure is Aphrodite, the goddess of love, beauty, and fertility. According to Homer, "her enchantment came from this: allurement of the eyes, hunger of longing, and the touch of lips that steals all widom from the coolest of man." With her fullness of passion and knowledge, with her expectations, memory, and mental present we take her to represent nootemporality, the complex time of humans.

The winged youth is Eros, in late Greek mythology the son of Aphrodite. According to Hesiod he was a primeval god who transformed Chaos so that the world might be created. To the lyric poets his concrete form was rain that brought heaven and earth into a creative embrace. For Parmenides he was the power that held together the contradictory aspects of the world. In Platonic cosmogony it was the principle of Eros, the love of the eternal, moved by visions of beauty, that conceived and bore "wisdom and all her sister virtues."

We take Eros to symbolize the open-ended hierarchy of unresolvable creative conflicts which arise from chaos in the course of evolutionary development.

Time as Conflict

A Scientific and Humanistic Study

J.T. Fraser

In experimental philosophy we are
to look upon propositions inferred
by general induction from phenomena
as accurately or very nearly true ...
till such time as other phenomena
occur by which they may either be
made more accurate, or liable to
exceptions.

Isaac Newton, *Principia*

The light, the light,
the seeking, the searching,
in chaos, in chaos.

From the conception the increase,
From the increase the thought,
From the thought the remembrance,
From the remembrance the consciousness,
From the consciousness the desire.

Maori Creation Myth

Birkhäuser Verlag   Basel and Stuttgart

CIP-Kurztitelaufnahme der Deutschen Bibliothek

Fraser, Julius T.
Time as conflict: a scientif. and humanist. study. –
1.Aufl. – Basel, Stuttgart: Birkhäuser, 1978.
  (Wissenschaft und Kultur; Bd. 35)
  ISBN 3-7643-0950-4

© Birkhäuser Verlag Basel, 1978
ISBN 3-7643-0950-4
Wissenschaft und Kultur, Bd.35
(Science and Civilization, Vol.35)

Made and printed by Birkhäuser AG, Basel
Designed by J.T. Fraser
Layout by Albert Gomm swb/asg

For my lovely wife
who goes to school every morning
and teaches little children
how to read and write
so that I may write and read
at home

Acknowledgments

For the gift of constructive criticism and encouragement, I am deeply indebted to a small group of friends: F.C. Haber, Nathaniel Lawrence, G.H. Müller, David Park, Lewis Rowell, and Georges Schaltenbrand.

The manuscript was edited by Susan Myers Felsenthal, average reader extraordinary.

It would have been impossible to complete this book without the continued hospitality of the Library of Manhattanville College and of the Burndy Library.

Credit lines for the illustrations will be found on the appropriate pages.

Contents

Introduction

THE THEORY OF TIME AS CONFLICT is a class of principles in the scientific and humanistic study of time. Its concerns include the abstract time of mathematical physics; time manifest in life and organic evolution; time relevant to man's behavior, collective actions, and conscious experience; and time that informs literary and artistic creativity.

In its mode of thought the theory follows the instructions of Isaac Newton for the development of scientific knowledge, but it also allows for the irreducible manifestations of the passionate faculties of man. The quotation from the *Principia* and the one from a Maori creation myth, both shown on the title page, represent two aspects of the human experience which the theory of time as conflict tries to accommodate.

The basic postulate of the theory is that time is a hierarchical structure of temporalities, and not a single one-way thrust. As we explore the distinct temporal levels of nature and inquire into their mutual relationships, we shall discover a hierarchy of languages and causations. This discovery when taken together with the basic postulate and examined in depth, leads our thoughts along paths toward the solution of certain problems which have remained intractable to conventional approaches. We shall also come upon some interesting questions never before asked.

This theory claims usefulness to fields of learning as different as genetics and physical cosmology, anthropology and aesthetics, sociology and logic. With such broad claims it ought not assume more than, but dare not assume less than: (1) a firm belief that the informed mind is capable of erecting universal structures of thought and (2) that it can do so in our epoch of proliferating knowledge, no less than in past and simpler epochs.

Few great ideas have been born through deductive reasoning. The nature of inventive thought is such that it usually permits

the establishment of continuity between the new and the old only
ex post facto, if at all. A class of new principles is not "there" to be
found as one might find the source of a river. Intellectual discovery tends to be a leap; it is not at all irrational, but neither is it
forced by reason. General principles seem to emerge from those
regions of scientific and humanistic knowledge where vision must
implement sight.

As an exploration of a fundamentally new idea, this book
contains reasoning based on established facts, speculations sus-
ceptible to experimental proof, and guesses which are subject only
to demands of self-consistency and predictive validity.

In the rich seventeenth century prose of Thomas Browne,
the creative element of the world functions "according to the
ordainer of order and the mystical Mathematicks of the City of
Heaven." But a working scholar, claiming no residence in
Browne's Heavenly City but only in the troubled villages of this
earth, cannot hope to formulate useful theorems unless he is
aware of and has examined in detail the issues which his thought
intends to accommodate.

I have done this in a recent work where I marshaled
whatever material I judged relevant and minimally necessary for
a comprehensive study of time. In that book [Fraser (1975)] I
followed the established method of argumentation. I reasoned for
and against ideas as they came up for consideration and supplied
the necessary notes and references. In the same volume I outlined
the theory of time as conflict.

This book begins where the prior work concludes. After a
chapter of summary and background, the theory is applied to
issues of interest to the working scientist and humanist. We begin
with physics, then go on to biology, the sciences of the mind,
cosmology, philosophy, epistemology, to the study of collective
behavior, to ethics, and conclude by considering the arts and
letters. In broad lines, the progression is from the sciences toward
the humanities, that is from the easy to the difficult.

This last remark calls for comments. The human condition
is familiar to all, one may argue, for we all live, love, suffer, and
die; scientific knowledge, however, may be acquired only through
years of concentrated learning.

It is true that scientific issues are often complicated and are best stated in special jargon, unintelligible to the uninitiated. But the thought processes necessary for scientific work are established in their form, if not in their content, and the rules of inquiry are strictly ritualized. Mathematization, for instance, is the most rigorous way we know to eliminate surprises. And underneath all quantified knowledge there is a belief in immutable, and hence discoverable, final truths.

In stark contrast to the regularity of Platonic forms, man is unpredictable in his action and thought. The central issues of individual and collective behavior are so far undefined; the course we should take toward an understanding and mastering of the malaise of our epoch, or of all of history, is unclear. The human predicament is so profoundly complex that, in a very non-mystical sense, it is unfathomable.

The last chapter of the book is comprised of some 300 problems. They offer an unguided but not whimsical tour through the study of time in general, and through the theory of time as conflict, in particular. Those who wish to develop courses or design seminars in the study of time, with or without emphasis on a particular field of specialization, may find this unique and rich collection of question both interesting and useful.

The power of the theory of time as conflict cannot be appreciated without an examination of its significance to a large number of issues. Therefore, partial reading of this work is unlikely to supply sufficient material for constructive criticism. However, the book might be read along either of two different tracks. Those conversant with scientific issues in the study of time may wish to proceed from beginning to end. Those who feel before they think, and agree with Rousseau that this is the common lot of humanity, might prefer to read the first chapter, scan two, three and four, and return to them after they have finished reading the balance of the work.

The book has no numbered references and only a few footnotes. However, a bibliography of the general sources used in this study is appended, together with a list of the literature cited in the text, identified by author's name and year of publication.

As the reader progresses along the path that leads to an understanding of the theory of time as conflict, the need for, and the reasonableness of this intellectual adventure will become, I believe, increasingly evident.

I   The Time and Times of Matter, Animals, and Man

THIS CHAPTER SKETCHES SOME OF THE CONSIDERATIONS that led to the formulation of the theory of time as conflict, outlines the major features of that theory, articulates its assumptions, and defines its vocabulary.

1   Search for Meaning and Order

Homo sapiens, man the knower, reflects upon the facts of passing time and, inspired and awed, searches for value, truth, wisdom, beauty and abstract knowledge. Homo faber, man the maker, organizes his daily activities with the assistance of natural processes which he regards as regular and predictable. The clockmaker and the clockwatcher are alter egos of the same organism, represented by the reader (and the writer) of this book. I shall sketch some of the activities of these alter egos by following first a philosophical trajectory, and then what might be called the clockmaking trajectory.

The Philosophical Trajectory

Two distinct views dominate the large body of thought concerning the nature of time. In their polarized extremes they hold, respectively, that ultimate reality is pure change or becoming, or else, pure permanence, or being. The mutual exclusiveness that characterizes the abstract motions of time as being or becoming also informs many parts of time-related concepts. *Necessity,* for instance, signifies events which must come about while *contingency* describes conditions which may or may not happen; *one* is that which remains identical with itself whereas *many* permits change; *infinity* need not ever show change while *finity* implies coming-into-being or going-out-of being; *nomothetic* or lawlike processes in nature imply predictability, whereas speaking of the *creative* in nature implies unpredictability.

These and similar pairs of concepts may sometimes be easily identified, sometimes only with great difficulty. But, although such identifications may make good working sense, a shrewd and well versed intellect can usually make any specific example of time-as-being or time-as-becoming look suspect. I could not discover compelling reasons why either class of ideas ought to be identified with time exclusively, to the detriment of the other class. Temporality subsumes both. The suggestion which emerges is that pure contingency and pure necessity are ideas of metaphysical status, justified only by usefulness in analysis. The Law of Contradiction (the postulate that a proposition and its negation cannot be simultaneously true) does not seem to apply to examples that deal with attributes of time.

I wish to postulate that the origins of temporality are neither in the nomothetic nor in the generative aspects of nature alone, nor in their possible coincidence as contraries, but rather in the conflicting separateness of certain opposites. I perceive these (yet unspecified) conflicts as examples of a universal condition of nature that I shall name *existential* (or *constitutive*) *tension*. In the ontology of time as conflict only existential tension may be identified with ultimate reality, but not any specific pair or pairs of opposites which we may recognize in the nature of time. It will be one purpose of this book to identify some such conflicts.

The Clockmaking Trajectory

Long before Homo sapiens came to interpret his experience of time in terms of such notions as finity/infinity, or necessity/contingency, Homo faber learned to organize his activities with the help of processes which, by the changing lights of various epochs, he judged to be predictable regularities. The story of this impressive skill is that of calendars and clocks. Their history demonstrates the successive reduction of ideas about regularities that involve time, to dynamic or static models (to wit, clocks and calendars) which represent those ideas to different but acceptable degrees of approximation.

While Plato projected man upward by perceiving ultimate

reality in the perfect laws of the cosmos, of which our local regularities were but bastardized projections, clockmakers retained man here on earth by reducing some of the ideal forms into useful models. Plato's cosmos worked in perfect harmony but the world of timekeepers never did. No timekeeper can be expected to reach the perfection which the laws, whose workings the timekeeper embodies, are believed to possess. It follows that identifying the idea of time by reducing it to any specific process cannot be expected to succeed.

We shall see later that time measurement necessarily involves the reading of at least two clocks, and that the accuracy of a clock is judged in terms of the reliability we can assign to the lawfulness of its workings. Regardless of accuracy, however, the ticking of clocks involves uncertainty about the future. For, although pine cones open regularly once a year, geese migrate predictably over James Bay twice a year, and cesium atoms vibrate reliably some 3×10^{17} times per sidereal year, there is no assurance that the next tick of the pine cone-, goose-, or cesium clock will surely happen. Here we reencounter the issue of lawfulness and contingency of the philosophical orbit, embedded in the practical concern of the clockmaker.

Whether we approach the universe as sages or tinkers, whether we seek temporal order through measurement or else a meaning to temporal passage, we seem to come upon the same conflicting separateness of certain opposites. Underneath our technical skills, as well as beneath our reflective modes of dealing with time we may suspect, therefore, the presence of the existential tension of man clockmaker and clockwatcher.

2   The Extended Umwelt Principle

The theory of time as conflict makes use of a principle first formulated over half a century ago by the German biologist Jakob von Uexküll. He noted that an animal's receptors determine its world of possible stimuli, its effectors its world of possible actions, hence, the world of each animal is determined by the potential functions of the totality of its effectors and receptors. He called this combined world-as-perceived and world-as-acted-upon the animal's *Umwelt*.* Events and objects that are not within an animal's Umwelt must be understood as not to exist in its world. If animals could write philosophy, poetry, or physics, their languages would be void of certain words that describe features which we regard as essential to the constitution of the world. Also, they would contain statements about events and things which we might not easily recognize.

The idea of Umwelts may be easily formulated for special sensory systems: one may think of auditory, tactile, and olfactory worlds. In complex organisms these Umwelts coexist and influence one another, yet they remain recognizably distinct. It is a logical step to go from special sensory systems to scientific instruments which, as has often been stressed, are exosomatic, that is, extensions of our (endosomatic) sensory systems. Thus, the Umwelt of a radio telescope is electromagnetic radiation, limited perhaps to the 21 cm. line of hydrogen gas; the telescope is blind to the light of the moon and deaf to the sounds of barking dogs.

I whish to go even further. Consider that often, after thorough inquiry, a domain of nature is revealed to us consistently and exclusively through a certain type of mathematical formalism and also that this formalism often cannot accommodate some features of the world which we ordinarily take for granted (or else, familiar features of the world appear in the equations in strange guise). That domain of nature to which such conditions apply might be regarded as a distinct Umwelt.

*The word *Umwelt*, in the context of Uexküll's work, has been translated as "self-world" and "phenomenal world." I find both renderings awkward. In earlier writings I translated it variously as "perceptive universe" and "species-specific universe." These, I believe, are better, but too long. The ordinary translation is "environment." But "environment" (in German, *Umgebung*) signifies an essentially unlimited, open region whereas *Umwelt* is delimited. Cf. "*Umwelt*: the circumscribed portion of the environment that is meaningful and effective for a given species and that changes its significance in accordance with the mood operative at a given moment." (English and English). For brevity and clarity I have retained the use of the German world, capitalized, but not italicized.

Uexküll's idea of species-specific worlds is not without relatives in the history of thought. C.S. Peirce in his doubt-belief theory of inquiry called those habits of behavior which an organism adapts for survival its pragmatic beliefs. The aggregates of all such beliefs, in Uexküll's terms, would belong in the animal's Umwelt. The idea is also Kantian in that it acknowledges that empirical reality cannot go beyond certain boundaries of existence. However, it is quite non-Kantian in that it admits many and distinct realities that make up an open set which might expand in the course of evolution, through the addition of increasingly sophisticated Umwelts. Going even further back in history, Aquinas held that "whatever is known is known according to the manner of the knower."

The whole, greatly enlarged scheme of the originally biological idea of species-specific universes I call the *extended Umwelt principle.*

3   The Hierarchy of Integrative Levels

The theory of time as conflict recognizes in nature a number of semi-autonomous integrative levels. They are successive forms of order that differ in their complexity, organization, and relative independence. The initial argument for a hierarchy of integrative levels is that they make good practical sense as do, for instance, Shakespeare's seven ages of man. As the argument of this book progresses we shall be able to refine this pragmatic view and point to epistemological differences among the integrative levels and hence, by the extended Umwelt principle, to ontological differences. We shall also inquire into the means and methods used by nature to keep them autonomous, by exploring the common boundaries between adjacent levels.

The basic substratum of the world is the universe of particles of zero restmass, traveling with the speed of light. For reasons that we shall learn later, we may call this integrative level the universe of first signals. Its model is that of an idealized relativistic gas made up entirely of photons, neutrinos, and gravitons (if they exist). Above this level is the world of particles with nonzero restmass. The model of this universe is that of a pure monatomic, non-relativistic gas made up of countable but indistinguishable particles. Above that level we find the astronomical universe of aggregate matter bunched into masses which form the stars, the star clusters, the galaxies, and the galactic clusters. The model of this world is that of heavy, cosmological particles distributed in the substratum of first signals and tenuous gas.

On a spherical object near one of the stars we recognize an integrative level of organic matter, comprising the cyclic and the linear (aging) orders of life. Out of many organic forms a single one evolved into human life, characterized by a highly developed central nervous system, including a very complex brain. I shall describe the integrative system whose hallmark is the mental activity of the human brain as noetic. Above the noetic integrative level we find the societal one, comprised of the collective work of human minds.

We have listed six major integrative levels: that of first signals, of particulate matter, of the astronomical universe, of life, mind, and society. In the vocabulary of the theory of time as conflict, each of these integrative levels is said to determine its distinct Umwelt. Each of these Umwelts supports a peculiar temporality, with each temporality subsuming the temporalities of the Umwelts beneath it.

4   Temporalities

So as to remove our attention from the many useful but misleading metaphors of time, we shall be talking not about time but about the levels of time, or *temporalities*. In this section I will list and describe briefly the temporalities associated with the integrative levels which were just discussed. We will begin with the simplest of the temporal Umwelts and work our way up to the temporalities of mind and society. Our understanding of each temporal Umwelt will grow as we critically examine various issues throughout the rest of this book.

There are processes which determine Umwelts that must be regarded as *atemporal*. By this is meant that there are no means, even in principle, whereby time can be recognized if such processes are examined entirely from within. In these worlds nothing can correspond to the idea of event, to conditions of before/after, to future/past/present, to causal connections, or to beginnings and endings. The foremost example of atemporality is the world of first signals, but we shall also encounter examples of atemporality among physiological processes, among the perceptive functions of animals and man, and in conscious experience.

There are processes which determine Umwelts in which time and space are distinguishable, even though events and things are often interchangeable. Events and positions may be specified only statistically and causation is probabilistic. I have named such primitive conditions *prototemporal* (for proto-, the first or lowest of a series). In prototemporal Umwelts our ideas of here versus there and now versus then are only loosely applicable. The inhabitants of prototemporal Umwelts are countable but they are not orderable by number. The foremost example of prototemporality is the world of non-relativistic particles, but we shall also encounter examples of prototemporality among physiological processes, among the perceptive functions of animals and man, and in conscious experience.

There are processes which determine Umwelts with temporalities of pure succession, that is, of succession without a preferred direction. In such Umwelts two events need not happen at once, yet there is no way of telling which event came first. Time may be said to flow, but past-present-future cannot be distinguished from future-present-past. I have called such Umwelts *eotemporal* implying the dawn of time (for Eos, Goddess of Dawn). Restating what has just been said, in eotemporal Umwelts events may be distinguished and identified as well as arranged by sequence, but only succession has any meaning. This kind of time may be compared to a series of blazes on trees, indicating a trail, but not to an arrow, indicating a preferred direction. Endings and beginnings of processes can be identified but they cannot be distinguished one from another; they both ought to be described by the same word – which we do not have, since our conscious experience of time is not eotemporal. Causation here is deterministic; the dynamics of this world is that of action and reaction. The foremost representative of eotemporal Umwelts is the Newtonian universe of solid astronomical bodies. However, we shall encounter eotemporal conditions among the organic functions of life, among the perceptive functions of animals and man, and also in conscious experience.

Living matter determines its own peculiar Umwelt, the *biotemporal*. It is only here that presentness becomes definable (in terms of the coordination necessary to maintain the autonomy of living organisms). As we progress from the simple and cyclic to the complex and aging order of life we observe a series of nows, or "creature presents," that differ in their quality and scope. Futurity and pastness become increasingly polarized, beginnings and endings increasingly asymmetrical, and the action-reaction connectivities of the eotemporal open up to final-, as well as multiple causation.

The Umwelt of the mind is *nootemporal*. Beginnings and endings are well defined and form the bases of private cosmologies whose central theme is personal identity. The "mental present" of man is lodged between his future and past. Its boundaries are ill-defined, for its contents varies continuously with the changing scope of expectation and memory. This is the Umwelt of the symbolic transformation of experience into signals and signs through which man communicates his thoughts and feelings. The connectivities of lower Umwelts are enlarged to include the functions of free will. This is the world often described as that of human time, or the arrow of time, or of time asymmetry.*

Above the nootemporal world we find the integrative level of cultures and civilizations. I shall argue later that for a global

---

*I shall refrain from using "time asymmetry" because, in spite of its popularity in the philosophy of physics, I judge it to be a misleading, and hence useless term. Consider the following quote from P. C. W. Davies.

Although we are forced to conclude that the laws of physics do not themselves provide a time asymmetry, it is one of the most fundamental aspects of our experience that as a *matter of fact*, the world is asymmetric in collective of man we may speculatively postulate a *sociotemporal* Umwelt. For reasons that we shall discuss, it is difficult to discern the hallmarks of the time, language, and lawfulness of this (potential) integrative level.

All temporalities below the societal are represented in the body and mind of man. Our biochemistry and our genetic machinery is either very close to or identical with those of all other living things on earth, even though the soma does not retain faithfully the record of its morphogenetic history. Our minds, however, do, and hence our mental faculties comprise many phylogenetically recent and many rather archaic functions. Accordingly, the theory of time as conflict assumes that the human mind, beyond displaying its peculiar nootemporal features, also subsumes functions that are appropriate to the lower temporal Umwelts. It would follow, then, that as we enlarge our knowledge of the integrative levels below the noetic, we should be able to recognize in their temporalities, causations, and languages, some features of our own conscious experience. This, I believe, may indeed be done both by intellectual and by artistic cognition (and, as is usually the case, by their mixture). However, imaginative sharing of lower temporalities is not easy; we are hampered by what I have called the *difficulty of regressive sharing*. The reasons

---

time. This is sometimes expressed by saying that the temporal asymmetry is "fact like" rather than "law like" or "extrinsic" rather than "intrinsic."

The use represented by this quote identifies "time asymmetry" with our temporal passage from cradle to grave; vide "fundamental aspects of our experience." In contrast the "law like" statements of physics are said to describe a timeless world, for they do not provide a time asymmetry. But nothing in nature corresponds to this simple dichotomy.

Thus, the eotemporal world is one of pure succession, hence of "time asymmetry," but in both directions. The biotemporal world is also one of "time asymmetry," but since life defines a "now," it makes possible the division of time into future and past. Finally, only in the nootemporal world can one begin to talk about human experience.

The absence of this fine structure from the current philosophy of science is unfortunate. By analogy, labeling all non-Americans as foreigners is a good enough approximation for certain limited purposes. But the same practice could hardly satisfy the anthropologist studying the human species.

for this difficulty stem from those functions of the mind which endeavor to secure the integrity of the self.

There is enough evidence to suggest that the sense of time in man (the conscious experience of time that determines the nootemporal Umwelt) is the outcome of phylogeny, that is, of the evolutionary history, which passed through steps that correspond to the hierarchy of temporal Umwelts; also, that these steps are recapitulated in their main features in the ontogenic development of the child. It is thus that the mind of man can be said to subsume all lower temporalities and possess the capacity for cognitive exploration of these Umwelts.

5   Laws of Nature

There are good reasons to believe that each of the major integrative levels of nature must be described in different, level-specific languages. By language is meant a class of signals and symbols in which the laws and regularities of nature must be expressed so as to satisfy the critical and practical intelligence of man, the formulator and tester of those laws. When the regularities and laws of the integrative levels are appropriately expressed, they too are found to be level-specific and hierarchical. To wit, each level-specific language and body of laws subsumes the languages and laws beneath it. However, the laws of each integrative level leave certain regions of their world unrecognized, hence *undetermined*. It is from the undetermined regions (seen ex post facto) that the regularities of the next higher level arise in a *nomogenesis*, or birth of laws of nature. It follows from these considerations that level-specific laws are unpredictable from Umwelts lower than the one to which they apply, and are inexpressible in terms of lower languages; also that no level-specific law or regularity may contain specifications in violation of lower order regularities: cats reproduce while stones do not, but cats and stones fall at identical rates.

The idea of nomogenesis must be sharply distinguished from suggestions that the laws of nature may be changing. In the epistemology of time as conflict, the laws of nature cannot be conceived of as changing; if and when they are found inaccurate, new laws are formulated. Nomogenesis claims only that regularities cannot be said to exist before the integrative level to which they apply actually exists: the circadian clocks of rats cannot be tested in the primordial fireball.

We shall see that undeterminism closely relates to the hierarchy of connectivities. To wit, we progress from instantaneous connections in the atemporal Umwelt to probabilistic causation, to deterministic causation, to final causation, to free will, and finally to historic causation. But undeterminism is not to be confused with *uncertainty principles* which arise when we try to specify the happenings of a lower Umwelt in a language appropriate to a higher Umwelt, when the descriptive language is too rich for the phenomena to be described. Neither should undeterminism be confused with *indeterminism* which is the property of complex systems, manifest in their incapacity to make accurate self-predictions.

6   The Strategy of Existence

The theory of time as conflict identifies each of the major integrative levels with certain unresolvable conflicts. By conflict is meant the coexistence of two opposing trends, regularities, or groups of laws, in terms of which the processes and the structures of the integrative level may be explicable. By unresolvable is meant that, by means indigenous to an integrative level, its conflicts may only be maintained (and thereby the continued integrity of the level secured) or else eliminated (and thereby the level collapsed into the one beneath it). If the conflicts vanish, so does the integrative level. However, the unresolvable conflicts of each level can and do provide the motive force for the emergence of a new level. But a new Umwelt, from its very inception, may once again be identified with certain unresolvable conflicts of its own. As the book progresses we shall describe the major conflicts of the different integrative levels.

How can conflicts be identified with integrative levels of nature? The wording is metaphysical, appropriate to universal theories. Yet the underlying idea is neither mysterious nor even nebulous but clearly isomorphic with Newton's famous third law of motion. "To every action there is always opposed an equal reaction."

Action and reaction cannot be thought of as existing independently. They are aspects of the same, single phenomenon: the state of motion of a body (including rest). But it is advantageous to think of single states of motion as produced by two opposing forces, because this analytical trick assists in the later development of our understanding of the physical world.

The opposing trends that the theory of time as conflict identifies with various levels of nature are also aspects of single integrative states; they cannot be thought of as existing independently. It is only their conflict which should be regarded as real. But it is advantageous to think of integrative levels as the manifestations of certain conflicts, because this mode of reasoning is useful in furthering the understanding of many processes.

The class of principles sketched in this chapter and further elaborated in this book, in connection with specific issues of interest to the working scientist and humanist, together constitute the "theory of time as a hierarchy of unresolvable conflicts."

II Atemporal Worlds

THERE ARE CERTAIN PROCESSES IN NATURE whose Umwelts must be described as atemporal. In this chapter we shall identify atemporal conditions in the physical world, among physiological functions of living organisms, and in the perceptive structures of man. By a name boJrowed from theoretical physics, I shall call all atemporal Umwelts chronons. In this extended meaning chronons, as a class of phenomena, are the atoms of time.

1 Physical Chronons

The primal universe is imagined as one of highly compressed matter of very heavy particles, the likes of which probably have not survived into our own epoch. We begin with the event which some physical cosmologies identify as a singularity in the solutions of their equations, and which distantly resembles the idea of Creation in some narrative cosmologies. At that singularity in the presently favored models of the physical world, the universe began to expand. We shall assume that we understand what is meant by "expand", even though later we will see that it is a rather difficult concept. We shall also assume that talking about seconds, hours, or years of time after the beginning of the expansion makes sense, even though we shall learn later that it does not.

Thus, a few seconds after creation the temperature of the fireball dropped to 10^{12} or perhaps 10^{10} °K., while the heavy particles decayed into photons, gravitons and possibly neutrinos. These particles of zero restmass are distinguished from all other forms of matter by the fact that they are the only ones that can travel at the speed of light, and in fact, they always travel; theirs is a world of ceaseless motion.* I shall refer to this early, somewhat

*It has been debated whether photons really have zero restmass. The lowest proposed limit that might be substantiated by measurement is about 10^{-58} grams. It is known from quantum field theory that what is being quantized in

idealized state of the world as that of photon gas, or relativistic gas, or the world of first signals. For reasons that will become clear later, I shall regard the condition of this world as both the basic and the primal form of the universe, although, as I have just said, more complex forms of energy seem to have been present while the temperature was above 10^{13} °K., and when the universe was less than a second old.

Special relativity teaches that in the life of a particle traveling at the velocity of light all events are simultaneous. What we recognize, for instance, as events along the path of a photon are changes, recognized after they happened, from within Umwelts of higher temporalities – such as when a physicist interprets certain lines on a photograph. It follows that a universe of purely relativistic energy, containing only photons, gravitons, and neutrinos, as judged from the point of view of any of the traveling particles, is one single happening. Of course, photons have no opinions but we may have some on their behalf, through the extended Umwelt principle. Thus, a universe of particles traveling with the speed of light is the true Parmenidean One of physical cosmology. In terms of the theory of time as conflict, such a universe determines an atemporal Umwelt.

Atemporality is *not* a way of talking about nothingness, or non-existence. The idea is isomorphic with an empty set which is a set, albeit empty. And since it has no discernible structure, the idea must also be taken as isomorphic with the idea of continuum. In classical cosmology it would represent the primordial chaos. Particles with zero restmass do have all kinds of curious properties such as their spins and their handedness. What they do not have

---

an electromagnetic field is not a localized object but a mode of wave field, something spread out over all of space. This is what we would expect if the photon is truly atemporal: it must fill all available space simultaneously. But then it follows that its Compton wavelength (its wavelength as a matter wave) could not be over the distance known as the radius of the universe. This puts a lower limit on the restmass of photons, 10^{-64} gm. in our epoch, decreasing as the universe expands. Be that as it may, there are good physical reasons which require photons to have zero restmass and we leave the issue at that.

are conditions that could be identified with the hallmarks of time, such as future-past-present; it is for this reason that their Umwelts must be taken as atemporal.*

Particles with zero restmass, especially photons, make up most of the matter of the universe; the cosmos, seen by our mind's eye, is almost entirely light. Upon our very small and quite solid earth light is the foremost synchronizer of physiological clocks and also the foremost carrier of energy needed for life. Light, as a symbol, informs all civilizations; even the word "idea" derives from the Greek *idein,* "to see." Ironically, since our imaginary observer (necessarily massless) must always travel at the speed of light on the back of a frozen photon, the world of pure light, as an atemporal Umwelt, is totally black. In the beginning was darkness.

The fastest conceivable connection among events is instantaneity. Since in an atemporal Umwelt everything happens at once, it follows that no connection faster than that provided by light should be identifiable in the physical world. This is indeed the case: as seen from any of the higher Umwelts, light delimits the speed at which events may be causally connected. Given the immense wealth and size of the physical universe, and given that, in our own nootemporal world, light is perceived as a limiting but finite speed of motion, there may be identified in the physical world an unlimited variety of locally, rather than cosmologically, atemporal conditions, or simultaneities. I shall give a few examples.

The smallest meaningful distance known in physics is the radius of the classical electron, 3×10^{-13} cm. The largest meaningful distance is the "radius" of the universe, perhaps 5×10^{27} cm. It

---

*Considerations based on quantum principles reveal, write Misner, Thorne, and Wheeler, that space and time are valid for the physical world only in a classical approximation. "There is no space-time, there is no time, there is no before, there is no after. The question what happens 'next' is without meaning." This, I believe, is overstated for what we have is a hierarchy of temporalities. To the dangers of labeling all conditions below the nootemporal as "timeless," we shall subsequently return in different contexts.

seems to be impossible to give any physical meaning to periods
less than that necessary for light to cross the electron radius (about
10^{-23} sec.) or to periods longer than that necessary to cross the
radius of the universe (about 2×10^{17} sec.). These two numbers
are the boundaries of atemporal Umwelts. The word *chronon* has
often been used as the name of the lower limit of this region. I
shall extend the meaning of this word to designate any length of
time which may be identified as atemporal (by a conscious
observer). Thus, the atomic chronon and the cosmological chronon encompass the range of all possible chronons.

The atomic chronon also obtains as an approximate limit to
the accuracy of an idealized clock; it is also the period of the
cosmic photon, the highest known electromagnetic frequency.
One may imagine temporal subdivisions of the atomic chronon,
but nature does not offer experimental means whereby events
separated by less than about 10^{-23} seconds may be identified, and
no theoretical extrapolation can be made for periods less than
about this limit.

Instead of a sphere with the radius of an electron one may
think of spheres of different radii, such as that equal to the span of
my hand (23 cm), or that which can contain a man (90 cm), or the
distance between the earth and the moon (4×10^{10} cm), or
between the earth and the sun (1.5×10^{13} cm). Corresponding to
these spheres we have, in addition to the atomic and cosmic
chronons, a hand-span chronon (8×10^{-10} sec), a body chronon
(3×10^{-9} sec), a lunar chronon (1.26 sec) and a solar chronon
(500 sec).

These and other chronons one may wish to construct define
atemporal regions or simultaneities which, however, make sense
only by their contrast to higher temporalities. Thus, consider a
mature experimenter sending out a light signal at instant t_0. As far
as all future instants are concerned, only those regions of the
world which are within the expanding light sphere can have any
temporal structure. Those portions of the world that are on the
sphere must be regarded as simultaneous with t_0 because the
Umwelt of light is atemporal. Furthermore, the light sphere is but
a moving boundary to the rest of the world external to it (with

respect to the time that began, for the observer, at t_0). Since, as far as the experimenter goes, nothing in those external regions can display any of the hallmarks of time, the whole universe external to and delimited by the expanding light sphere is atemporal, by definition. We may now imagine expanding light spheres generated at equal intervals. For each instant in the experimenter's life those atemporal regions of the universe that are outside the light sphere (sent out at a reference instant) are those of the absolute elsewhere, familiar from special relativity theory.

The idea of absolute elsewhere follows from the celebrated Lorentz transformations of special relativity theory. It obtains as one explores the space-time character of that theory by means of imagining relative motions with speeds increasing from zero to above that of light. The issues are best introduced by means of a geometrical device that represents the Lorentz transformations for plane-time (rather than space-time), known as the Minkowski diagram. The following critique of relativistic time is given with the assistance of the Minkowski diagram, because visual representations are generally more easily grasped than algebraic ones. (See Figure 1b on p. 55).

The time axis of the Minkowski diagram is said to accommodate futurity, pastness, and presentness; therefore, it must represent nootemporality. As one traces the world-lines of points traveling at relative speeds from zero velocity to that of light, the famous time dilation of special relativity theory obtains. Finally, for the framework of a traveling photon, proper time vanishes. The overall impression one has when going through the appropriate purely mathematical exercises, is that the nootemporality of the time axis smoothly transforms into the atemporality of the light cone. But this view is misleading because the nootemporality represented by the time axis is separated from the atemporality of the light cone (and regions beyond) by several integrative levels of nature.

Working backward from the atemporal regions, before one is entitled to give the usual meaning of futurity, pastness and presentness to the time axis, one must first leave the atemporal Umwelt for the prototemporal one, then evolve through the eotemporal to the biotemporal one, and only then may one determine the Umwelt whose time the time axis is believed to represent.

No clear analysis of the Minkowski diagram is possible if the hierarchy of temporalities is neglected. It is not surprising that efforts to reconcile the meaning of the unstructured representation with reality produced a literature of repetitious confusion.

Turning now to the atomic and the cosmic chronons, we note that they differ not only numerically but, so it would seem, qualitatively. The atomic chronon is somehow atemporal "inside" whereas the cosmic chronon is atemporal "outside." We shall try to understand this inflection of viewpoints by electing to involve man, the observer.

Let us imagine that we receive light signals from the moon, the sun, and the edge of the universe. I shall know that these reports are from my past and that the respective bodies have already been in my atemporal Umwelt for 1.26 sec., 500 sec., and 10^{17} sec. Furthermore, they will remain in my atemporal Umwelt until such time that I can begin interfering with their fates 1.26 sec., 500 sec. or 10^{17} sec. hence.

Furthermore, as I watch the light signal arrive from the moon, the sun, or at the edge of the world, I will have to allow activation times for the prototemporal functions of the molecules of my retina, then for the signals to reach my brain, and then for me to take conscious action in terms of my store of memories and expectations. The same argument holds in reverse for sending out a signal. A conscious decision appropriate for the nootemporal Umwelt must work its way across the lower Umwelts before a message may be sent to the moon.

The average reader of this book would fit in a sphere having a diameter of 180 cm. The corresponding physical chronon is about 3×10^{-9} sec. Along the surface of his skin the maximum speed of communication drops seven orders of magnitude, to about 5×10^3 cm/sec. Within the body the limiting velocities are the rates of propagation of changes important for life, such as the transfer of metabolites, pH changes, and the like. It is even possible to define regions of meaningful absolute elsewheres for physiological processes within an organism, as has been done for embryogenesis by Brian Goodwin. Within the skinbag, the physical chronon joins a class of biological chronons and, as we shall see, even a perceptual chronon of about 2×10^{-3} sec.

Because of the substantial changes of speed that determines regions of absolute elsewhere for living processes, there is an inflection in the position of chronons that is shorter than about 10^{-3} sec., as far as our experience goes.

Consider that I could send a light signal from my thumb to my little finger and say that the little finger is incommunicado to the thumb for 10^{-10} sec. In reality, it is incommunicado for a period that may be ten orders of magnitude longer. I can await the distant sun, or Mars, or the edge of the universe to emerge from the atemporal physical world, cross my lower temporal Umwelts and make itself manifest to my consciousness. The situation for body-, hand- or atomic chronons is inverted. We can only reconstruct them; we cannot await their emergence from the atemporal matrix. Accordingly, their existence must be assigned to an internal world. Indeed, short chronons appear to us as though beheld from an external vantage point, whereas long chronons are seen as though beheld from an internal vantage point.

But the method of construction of physical chronons remains the same from electron to cosmos: they are all lengths of time (as judged by us) needed for particles with zero restmass to cover certain distances. Hence, physical chronons form a continuous spectrum; the inflection of our viewpoint is a result of the intrinsic speeds of our bodily functions combined with the hierarchical character of time.

Atemporal conditions cannot be placed end-to-end to add up to higher temporalities. The Umwelts of chronons are atemporal; hence they have no ends, beginnings, or middles. If the man on the sun sends me a light signal which I immediately relay to the full moon, then I have constructed a sun-full moon chronon which, however, is as atemporal as either of its portions. To get away from atemporal conditions we must advance to new integrative levels.

According to the Copenhagen interpretation of quantum theory, wave and particle representations must be regarded as identically valid, as in the case of photons as waves and photons as particles. Historically, the wave properties of light were known first and when its particle properties came to be studied they were expressed in terms of wave properties; the reverse holds for matter. Thus, by mental inertia, we tend to associate light with waves and matter with particles. But these preferences are irrelevant.

What is important, however, is that the interchangeability of wave and particle representations neither guarantees the practicability of, nor demands that equal emphasis be placed on both the continuous and discrete representations across the hierarchy of physical reality. As we deal with the evolving universe, or else with structural forms in the present universe which correspond to early developmental steps, we must use whatever representations are best suited for the phenomena at hand. For instance, in microphysics if we wish to explain experimental results, we are well advised to talk about interactions rather than attributes of individual particles such as mass, charge, energy, or momentum. As we approach massive systems, the consideration of attributes becomes methodologically permissible and, finally, necessary.

Thinking of particles with zero restmass, we are entitled to form a mental image of an atemporal world, filled if you wish, with waves of light. As we begin considering intergalactic dust, the wave representation of light and matter does not become untrue but does become impractical: the preferred modes of description are those of statistics, magneto-hydrodynamics, chemistry. Again, as we leave the prototemporal world of gas, dust, and plasma and begin dealing with aggregate matter, the former representations do not become untrue but do become impractical. A three-fold division of the physical world implied by these thoughts is also suggested by the distinct foci of the three great theories of physics.

Special relativity theory is essentially a theory of the atemporal Umwelt: it ties motion, mass, and energy to the world of light. Quantum theory is most useful in dealing with the prototemporal world, whereas general relativity theory has its major significance for the eotemporal world of vast, gravitating masses. Although all three theories interact, partially overlap, and necessarily refer to the same world, they do emphasize such aspects of that world which the theory of time as conflict identifies with different integrative levels. The profound difficulties of combining quantum theory with the complete machinery of relativity theory are well known.

By the extended Umwelt principle, which suggests that epistemological facts be identified with ontic status, the technical division implicit in the foregoing discussion must be interpreted to mean that in the physical world we are dealing with three distinct integrative levels – each with its own temporality.

2   Physiological and Perceptual Chronons

Let a human subject determine through listening whether two brief soundbursts are distinct or not and, if distinct, what is their sequence.

There is ample evidence that if the separation of the signals is about 2 msec. or less, they will appear to the listener as one single sound. The actual numerical value is a function of the amplitudes of the stimuli, the perceptual modalities used, and of other variables. The upper boundaries of this interval are sometimes called the fusion threshold. We have identified perceptual chronons, sometimes called "chinks," sometimes "quanta," because within this period, in the Umwelt of the subject's hearing apparatus, it is impossible to identify any of the hallmarks of time. How many signals we "really" have within the chronon we know through instruments which change the temporal separation between the signals into visual or other suitable displays, so that the distinctness of the signals can become perceivable. For audio signals, tactile inputs, and for visual excitation the fusion limits are usually different but in all cases the limitations of our senses provide us with instances of atemporality among our perceptive functions.

The oilbird of tropical America uses echolocation for its nocturnal navigation. It sends out short pulses separated by 2 to 3 msec. and identifies the position of the echo between two outgoing signals. The bird must, therefore, be assumed to be able to discriminate conditions of before/after within periods of 2 to 3 msec., but surely its ability to do so ceases at some shorter interval. The perceptual chronon of the oilbird is, perhaps, 0.5 msec. The mustache bat of Panama uses a complex system of 30–60–90 KHz continuous wave and frequency modulated signals, with the FM signal being 2–3 msec. long. Somewhere in this rich spectrum there must be a mustache bat chronon.

Fighting fish are ill-tempered three inch animals that live mainly in the rivers of Northern India and Thailand. They are so combative that when confonted with a mirror they attack their own images. Through very elegant experimental techniques in which a stroboscopic disk with adjustable sectors was rotated at different rates between the fish and a mirror, it was shown that the animal has a visual fusion threshold of about 30 msec. This, then, is the perceptual chronon of *Betta splendens*. Experiments with dog movies suggest that the perceptual chronon of dogs is about 100 msec. It was first reported in 1932 that four or more tactile stimuli applied by a vibrating reed to the belly of a snail hanging by its shell, makes it extend itself, ready to crawl on the reed. What humans would judge as three or four separate events, seems to appear to the snail as one single simultaneous experience which implies a solid surface. The snail's fusion threshold, hence its perceptual chronon, is perhaps 300 msec.

Returning to human subjects (see Figure 1a), we note that two bursts of sound separated by more than about 2 msec. usually enable the listener to recognize two distinct signals but, as long as the separation is less than about 20 msec., he cannot tell which signal came first. Thus, above the fusion threshold we encounter conditions of time perception which permit the counting of events but not their arrangement by number. Such conditions are the hallmarks of a prototemporal Umwelt.*

If the period between two brief stimuli is more than a certain lower limit (anywhere between 20–50 msec. depending upon the subject's mood and on perceptual modality), a human subject can often not only identify the signals as distinct and count them, but he can determine their sequence correctly. These lower limits are sometimes called the order threshold. The possibility of arranging events by number is a hallmark of an eotemporal Umwelt.

Somewhere above the order threshold we enter a domain where it becomes possible to test for the capacity to estimate signal lengths. Interestingly, the ability to distinguish and order tactile, visual, or auditory stimuli is not necessarily the same as the ability to perceive signal lengths as equal to that actually generated by a source. The perception of stimulus equals the stimulus length only if that length is above a certain critical value; for visual excitations that is about 130 msec. If the stimulus length is shorter but still of sufficient intensity, the signal will be perceived as though equal to the critical length.

The neurological processes which determine this threshold value are obscure, but one may assume that they represent a minimal processing time of some sort. Although various experimenters have been using different criteria as to what they mean by the basic processing period, it is clear that above the eotemporal region of perception we observe a complexification of nervous processing that, in its most sophisticated form, makes it possible to insert a conscious decision between stimulus and action. I would like to postulate that between the eotemporal present and the shortest period in which conscious decision can be

---

*An example of the prototemporal present is encountered by all typists. In rapid typing the individual letters follow one another often within the period of the prototemporal present. Accordingly, while words are correctly spelled, individual letters are often transposed. As one ages, this phenomenon gets wroes.

TOPOLOGY OF THE MICROSTRUCTURE
OF HUMAN TIME PERCEPTION

On the abscissa of this graph we plot the activation times of a hierarchy of neurological responses, which, we assume, are responsible for the microstructure of time perception. Since we are interested only in the topology of the microstructure, the scale is arbitrary. The times on the abscissa would be measured in convenient units, such as milliseconds.

The ordinate accommodates those levels of temporality which we can identify in the microstructure of time perception. Corresponding to each temporality we define a present. In a conscious subject all the organismic responses, or tracking systems, function simultaneously. Therefore, the hierarchical components of the *now* are *nested presents*.

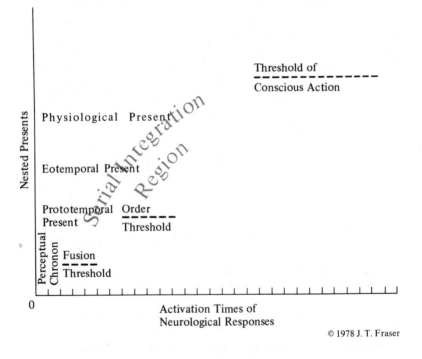

© 1978 J. T. Fraser

Fig. 1a Topology of the Microstructure of Time Perception

When brief stimuli are separated by very short intervals, they fuse into perceptual chronons. The lengths of the chronons depend on the perceptual modalities involved, on the amplitudes of the stimuli, and on other variables. The upper boundary of this region is the fusion threshold.

As stimuli become separated by longer intervals they become discriminable and hence, countable, but not necessarily orderable. Such conditions make for a prototemporal present. The upper boundary of this present is the order threshold.

As even longer periods are explored, we enter the serial integration region. By varying the pacings and the intensities of the stimuli, a number of different perceptual experiences may be produced. Though this region is ill-explored, it is nevertheless possible to identify conditions under which the subjects can not only count but also order the stimuli. Such conditions of pure succession make for an eotemporal present.

The ability to distinguish among stimuli and order them in sequences is not the same as the ability to perceive correctly the signal lengths generated by the source. If stimuli lengths are shorter than a certain minimal value but still of sufficient intensity, the signals will be perceived as though equal to that critical length. I postulate that this period represents yet another minimal processing time and I shall call it the physiological present.

Finally, we approach lengths of time which permit a subject to make a conscious decision in response to a stimulus. This length of time is the lower threshold of conscious action. Life processes that involve changes shorter than this limit must remain reflexive and unconscious.

Nothing in the diagram corresponds to the mental present because, as we shall see later, the mental present of man has no definable temporal limits.

made there is yet a distinct phase in the microstructure of human time. Since this region is between the eotemporal and the nootemporal Umwelts, hence of a biological nature, I would like to call it the *physiological present* of man. Living processes that are faster than the upper boundaries of this present must remain reflexive and unconscious.

In many animal species their (presumed) perceptual chronons open up and the atemporal conditions evolve into capacities for delayed gratification, anticipation, and even rudimentary ideation. It is reasonable to assume that with properly designed experiments, fusion thresholds could be identified in many species and that above such fusion thresholds other conditions may also be found which correspond to the various presents in man. However, because of a combination of reasons they remain incomplete compared with the sense of time in man. By a term whose meaning is left intentionally fuzzy, I shall call the highest present in animals their *creature present*.

In the course of ontogenetic development, the physiological present of the infant (unlike the creature present of animals) gives rise to the *mental present* of man. Whereas the presents that correspond to atemporal, proto-, eo-, and biotemporal conditions have definable boundaries, the mental present, as we shall see, has not.

The various traces of temporalities in the microstructure of our experience of time which we have so carefully distinguished are usually classed under the single designation of timelessness. This mistake is widely observable in the sciences as well as in the humanities. Its source is probably the fact that the lower temporalities, when compared with nootemporality, all appear to be rather incomplete. We shall discuss the experience of timelessness later; here we only warn against mistaking lower temporalities for the total absence of time.

To assist us in dealing with the many levels of temporalities represented in the psycho-biological organization of man, I shall distinguish *time perception* from the *sense of time*. Under time perception I will class those functions of man that call upon his capacities to function in any of the temporal Umwelts below the noetic, such as calling upon his perceptual motor skills related to brief time intervals, or upon his biological functions controlled by physiological clocks. The various thresholds identified earlier are associated with time perception. Under time sense I will subsume those aspects of human behavior and thought in which the symbolic transformation of experience is preeminent, such as in long term expectation, long term memory, personal identity, language, and the social structuring of time. Time perception and the sense of time blend into each other and sharp demarcation lines between them cannot usually be drawn, but the division makes good practical sense.

By now we have wandered away from the clear-cut examples of atemporality represented by physical chronons. This detour has given us at least some ideas about how the class of chronons is enlarged by adding physiological chronons to the physical ones and how, in their turn, physiological chronons are contiguous with Umwelts of more advanced temporalities in the perceptive faculties of man.

3   Chronons as Atoms of Time

The idea of atemporality is implicit in the ordinary use of the concept of "event." The meaning may be made explicit by defining event as any thing or condition that remains identical with itself through a period of time, in terms of the system which makes the determination. When so recognized events may be given names and may be quantified.

"One shot in the dark" as perceived by a listener is one event even though instrumental measurements may reveal, ex post facto, that there were two shots. "One particle emitted" as defined by a competent physicist is an event, as is "one night" or "one day" as understood in the Book of Genesis. While the event endures no variation is permitted in its character: "one night" ceases to be that if interrupted by the rising sun. "One life" is also an event as understood and declared by a midwife or an executioner, but it ceases to be that if interrupted by death. "My life" is the primary reference of simultaneity in the world of selfhood; within this single event "my birth" and "my death," from the point of view of "my life," cannot be separated.

In general, then, an event is a chronon and all chronons are single events; they are simultaneities or atemporal Umwelts without distinguishable inner temporal structures, such as succession or directed order.

Earlier we encountered atemporality as the Umwelt of

particles that travel at the speed of light. On best evidence there are some 10^8 times as many photons in the universe as all other particles taken together. Clearly, then, the atemporal physical Umwelt is the basic substratum of the world. In this atemporal sea of absolute restlessness we find vast regions of gas and dust: these determine prototemporal Umwelts. When observing the sky we see our own and millions of other galaxies. These objects are isolated eotemporal Umwelts. Though enormously large by human standards, they fill only one-millionth of the volume of the knowable universe, yet contain almost all of the matter of that universe. Regions of massive matter are truly few and far between. On one of these islands, on a diminutive speck of matter, we may further identify bio-, noo-, and socio-temporal integrative levels.

Our world of biological functions, our world of discourse, thought, and emotion is built from events which, as I have argued, are instants of atemporality. The class of physical chronons is extended by life, which adds to it an open set of physiological chronons; to these the mind adds all other identities that our psychological faculties can appreciate and our imagination can generate.

All higher temporalities are constructed from chronons whose relationships create the principles of the temporal hierarchy: countability, arrangeability, nowness, preferred direction, and so forth. Thus, chronons are the atoms of time. Here as elsewhere, atomism closely relates to the logic of explanation: what do we explain by what. In this case we explain temporalities by connecting in many and different ways the many and different atoms of time which, in themselves, are indivisible and continuous.

Our situation as humans is rather insular. It might be described as one of creative bursts, with the vapor of higher temporalities issuing from underwater volcanoes beneath a sea of atemporal, dark light.

III   The roots of time in the physical world

FROM PLATONIC IDEALISM UNTIL JAMESIAN PRAGMATISM, the intellectual winds of the West blew much, both hot and cold. The history of ideas is replete with headshaking and bloodshed, as man was moved from the center of the world to a small planet by the Copernican revolution, and from a station apart from nature into a position as a part of nature by the Darwinian revolution. Yet, from the privilege of retrospect, the world of physics from Aristotle to Newton may still be described as comfortable. By this I mean that the material which was to be explained, or explained away, was almost entirely that obtained by the biological, endosomatic senses of man. Only when the power of our native biological instruments came to be enlarged through the power of the exosomatic instruments made by a sophisticated artisanry, and only, with the cumulatively growing force of quantified knowledge (perhaps from the turn of the 18th century) could certain uncomfortable suspicions arise. Physics began to ask questions that would have been unthinkable to the best of classical scientific minds.

It has been repeated ad nauseam that this upheaval in physics changed our understanding of time. The ordinary interpretation identifies this change with such discoveries as the relativity of duration and of certain sequencing revealed by special relativity theory. I shall argue that these are important but tangential issues. The most significant insight of twentieth century physics is that relativity theory, quantum theory, and thermodynamics, when considered together, inform us of the hierarchy of causations and temporalities of the primitive Umwelts of nature. We shall find, however, that in spite of their universal power and stunning novelty, these revelations cannot elucidate the temporal issues of life and mind.

1   Cosmic, Absolute Unrest: the Universe of First Signals

We found earlier that a universe of particles with zero rest mass is a single atemporal event, a vast simultaneity, a true Parmenedian One, that remains identical with itself. Also, since nothing can be faster than instanteous, events in our experiential world of higher temporalities cannot be connected by any physical means that travel faster than the speed of light. For these reasons, the world of photons, gravitons, and neutrinos may be described as that of first signals. In a pleasing way, these particles may also be seen as the historically fundamental, first states of energy.

It is surprising to learn that in the atemporal world of pure oneness, or being, there is nothing that could correspond to our ideas of rest: there is no such thing as a non-moving, or even a sluggish photon. The atemporal world of first signals is one of absolute unrest, pure change, unspoiled Heraclitian becoming. In this curious double nature of pure being as well as pure becoming we might imagine the primordial unresolvable conflict of existence, the original constitutive tension that later communicated itself in many and specific ways to matter, beast and man.

In Newtonian kinematics, which is the study of motion in terms of velocities, a universal framework of absolute rest and absolute time were assumed. This is one of the comfortable beliefs that had to be rejected when relativity theory revealed that, if we are to obtain self-consistent physical laws to account for observed phenomena, (mainly those observed by means of sophisticated instruments), we must dispense with all reference to absolute rest and rebuild our ideas of motion with reference to constant unrest. The physical process identified as the new referent was the propagation of electromagnetic waves or, simply, that of light, if we use the name of the visible spectrum pars pro toto.

In the relativistic understanding of the physical world it is quite appropriate to call the motion of light absolute because its role in relativistic kinematics is isomorphic with that of absolute rest in Newtonian kinematics.

Consider that in the spirit of Newton's *Principia* one can legitimately speak of states of absolute rest, so recognized by all informed observers. Based on the daily experience of many millions of years, the idea of absolute rest corresponds to a truth of nature, to conditions that are self-evident. The statement that no meaning can be attached to speeds less than zero sounds rather silly: when my horse stops running it stands still; it cannot go slower than not to move at all. If one is conversant with Newtonian truth, a state of absolute rest should be obviously invariant with respect to all observers.

In relativity physics it is the speed of light which is invariant for all observers. That this is indeed the case is a truth of nature learned through the use of our exosomatic capacities. The statement that no meaning may be attached to speeds above c, the speed of light, sounds surprising for we have no experience of such high speeds. Had we, as a species, evolved with feet that could carry us at rates of 298,000 km/sec., then the existence of an upper limit to relative velocities would probably be self-evident. It would be judged to be a truth of nature that derives from the inevitable logic of reality. Its experiential consequences would be built into our perceptive faculties and incorporated in our languages. Our thinking would be so formed that any assertion that relative speeds larger than that of light cannot be found, would sound banal.*

According to Newton's celebrated first law of motion, a body will retain its state of translation unless compelled to change it by forces. Its absolute velocity should be calculable by all observers from different numerical values that such velocity would have, depending on the motion of the observers. Special relativity theory generalizes this law: a body of zero restmass will retain its state of translation, and its velocity will have the same numerical value for all observers, quite regardless of the observers' states of motion.

---

*Thomas Wright of Durham was neither a prophet nor a relativist, but creative thought has its way of being expressed in many dialects. Thus, writing in 1750, he remarked that "Vision, light, and electrical virtue, seem to be propagated with such velocity, that nothing but God can possible [sic] be the vehicle."

Newtonian rest and the motion of light may thus be seen as alike and symmetrical in their absoluteness; but they are quite unlike in their ontic status. Whereas the phenomena of electrodynamics and mechanics possess no properties that would correspond to the idea of absolute rest, and hence a framework of absolute rest does not seem to exist, we have found a universal framework that corresponds to the idea of absolute unrest both in mechanics and in electrodynamics. This framework is the motion of light. Its absoluteness is with respect to all conceivable motion; its uniqueness is its primordial atemporality. The security of absolute rest in a Sensorium Dei was thus replaced by the insecurity of the ceaseless unrest of a secular cosmos.

This revolutionary change of emphasis is embodied in special relativity theory which tells us how the absolute atemporality of light relates to the character of time on the different integrative levels of the physical world. Specifically, I want to consider the plasticity of the fundamental variable of that theory, which is the four-distance of space-time. A four-distance is the separation of two events in a combined representation of space and time. ("Time" and "event" in this definition are used in their ordinary, unexamined meanings).

Let us consider the lowest and highest levels of the physical world: the atemporal and the eotemporal Umwelts.

First, assume two reference events as happening in the life of a photon. The four-distance is zero, the temporal separation of the two events is a minimum, and we are in an atemporal world. Next, let us move to the vicinity of a massive inertial body and consider the same two events at relative (local) rest with respect to that body. Their separation may now be measured by a clock. But we still need the sense of time of the physicist because we are in an eotemporal world where time is succession without direction. The clockmaker has to supply the arrow of time, for there is nothing in relativity theory that could be used to define nowness, futurity, or pastness. Still, we can make sense of the separation of the events and express it in units of time. What we now have is the maximum possible separation of the two events in time.

Between the atemporal and the eotemporal regions of nature there is the prototemporal integrative level. Here particles may remain (on the average) at rest or may move at speeds approaching that of light. The time separating the two events may be any number between zero and the maximum, corresponding respectively to atemporal and eotemporal conditions. This plasticity of time is responsible for the famous time dilation of special relativity theory.*

The same theory also permits us the writing down of four-distances that correspond to connections among events, faster than the speed of light. In terms of our analysis such "space-like distances," as they are called, amount to reaching into the structure of the atemporal world. But since atemporal Umwelts have no temporal structure, nothing in nature can correspond to such four-distances and as far as we know, nothing does. In this frame of thought, motion faster than light corresponds to the suggestion of motion slower than rest.

2   The Imperfect Speaker: the Particulate Universe

For a few thousand years after the beginning of its expansion, the universe was dominated by radiation, but as the tem-

---

*Let the two reference events be the birth and death of a man. Send out a death ray at the speed of light upon his birth, reflect it from a mirror 10^{19} cm. away. As the light returns and hits the man, he dies. How old was he?

In the atemporal Umwelt of the photon the two events are simultaneous. In the chronon known as "my life," which is my private cosmology, the events are also simultaneous. However, you might elect to travel along with a not-too-fast particle leaving at the instant of birth and returning upon his death, and measure his life span as 21 years. Or, you may remain with him as a surviving friend and declare that he died at the age of 80.

It is the neglect of the hierarchical structure of nature that makes the clock problem of relativity theory difficult to integrate with the rest of our understanding of the world. It is not possible to transpose the pencil-and-paper clarity of the issue into nature because of reasons that are discussed i connection with Figure 1b. See p.55.

perature dropped below $\approx 10^4$ °K, an increasingly larger portion of the energy changed into forms of elementary particles with non-zero restmass. For our purposes the model of this universe is an idealized, non-relativistic, monatomic gas. The particles of such a gas are indistinguishable from one another, hence they may be counted but not arranged according to number; in the mathematics of such a world cardinality would predominate. These are some of the hallmarks of a prototemporal Umwelt.

Space and time in the prototemporal world of the particulate universe are distinguishable only weakly, because individual histories and group actions at an instance are interchangeable, as are things and events. Connectivity among events is that of controlled randomness, hence probability is the paradigm of physical law.

Probabilistic lawfulness is curious: it describes collective actions among individuals which, by definition, must be unconnected as well as without memory. Thus, for instance, by the famous ergodic theorem of statistical mechanics ensemble averages and time averages are interchangeable. *

The esprit de corps among members of a set which is not supposed to form an interconnected corps, the memory of members of a set which are not supposed to have a memory, or the identity of spatial and temporal distributions should not be attributed to subtle and hidden variables. On the contrary, the conditions suggest a very primitive and unsophisticated Umwelt. The laws of chance which rule the proto-temporal world do not hide our ignorance, they only proclaim the ultimate truth of an

*Take 10,000 identical dice and on 10,000 identical dice-throwing machines throw all of them at once. Record the numerals turned up; this is your ensemble. Take any one of the dice and on one of the machines throw it 10,000 times. Record the numerals turned up; this is your history. Compare the distribution of the ensemble and of the history. They should be very close, perhaps even identical.

How do 10,000 simultaneous but unconnected events communicate to produce a more or less predictable outcome? How does a dice without memory, trace out a predictable temporal behavior? How do these spatial and temporal distributions communicate to yield identical results?

integrative level. This view has been maintained in microphysics. Quantum theory conceives of statistical regularity as a brute and irreducible fact of nature. In the context of this chapter these ideas are extended as applicable to all prototemporal conditions, and attention is drawn to the weak separation of temporality and spatiality implicit in such a state of affairs.

Probability in science has a dual significance. It may describe a law of chance, or it may signify a degree of belief in the absence of thorough understanding. In the epistemology of the theory of time as conflict, causal connectivities in prototemporal Umwelts are seen as irreducibly aleatory. Only when they appear in regularities of the eotemporal Umwelt, are they seen as hiding our ignorance, because causality in that world is deterministic. For instance, as the laws of planetary motion became slowly clarified, statistical uncertainties in description and prediction became increasingly smaller (though not less interesting). The regularities of radioactive emission, however, were, and remained, probabilistic.

The prototemporal nature of quantum theory permits certain interpretations of experimental phenomena, such as would be unthinkable for the physics of the eotemporal world. For instance, the Feynmann interpretation of pair annihilation and production suggests that the positron be regarded as an ordinary electron traveling backward in time. This is a plausible interpretation of certain experimental facts, but it has been criticized as being no more than a way of speaking; and it is. But that the suggestion may be at all seriously entertained points to the symmetry of time in the prototemporal. It would be difficult to select embryogenetic processes, for instance, which would be describable as happening backward in time, even as a way of speaking.

In the prototemporal world of microphysics it is impossible to establish a well defined unit of time because no operational method is conceivable whereby a certain length of time could be reasonably preserved and transported. Prototemporal clocks are unpredictable and their temporal events sometimes turn into things. One is well advised not to even employ space-time as the framework in which microphysical processes go on: it is better to consider momenta, energy, and fields. For instance, interference phenomena of matter fields is visualized through probability wave amplitudes. From here one can proceed to calculating numbers of particles arriving in a small region of space within a unit of clock time. It is impossible to specify times and places of arrival of particles in any other way than through the tenuous means of relative probabilities.

Once upon a time on an English heath, two men encouraged three witches to tell the future of individual grains, out of a bushel of identical grains.

> BANQUO: If you can look into the seeds of time
> And say which grain will grow and which will not,
> Speak then to me ...
> FIRST WITCH: Hail!
> SECOND WITCH: Hail!
> THIRD WITCH: Hail!
> MACBETH: Stay, you imperfect speakers, tell me more ...

As were the witches, so is the probabilistic world of indistinguishable particles an imperfect speaker; it cannot yet say things more specifically than in statistical, depersonalized forms. Prototemporal causation displays an irreducible uncertainty.

3   Local Relative Rest: the Astonomical Universe

Some ten million years after the beginning of the expansion of the universe, when average radiation temperature dropped to about 1000 °K, the material of the prototemporal world began to coagulate into massive galaxies of solid masses. A not altogether metaphorical description would say that heat energy became frozen and, passim, the absolute unrest of the atemporal world became arrested. The collective properties of this arrested energy determine the nature of the eotemporal Umwelt of the physical world. Galaxies, which are the fundamental particles of the astronomical universe, are made up of matter in a virtually unlimited variety of forms, and hence unlike members of the prototemporal Umwelt, the members of the eotemporal physical world are not only countable but also distinguishable and therefore arrangable by number. The mathematics of an eotemporal Umwelt may therefore subsume cardinality as well as ordinality.

The astronomical world is immensely richer in potentialities and is, somehow, more interesting than the particulate world. In this section we will discuss some aspects of this great wealth.

Although, as we learned earlier, no universal frame can be experimentally identified that would correspond to our idea of absolute rest, it is possible to identify an infinite number of frames which correspond to our ideal of local, relative rest. They are called inertial systems and may be defined as those in which the temporal separation of any two events, happening at the same location, is maximum. When the two reference events are observed from another inertial frame in relative motion they will, of course, appear at different positions (with respect to the moving frame) and their temporal separation will be measured anywhere between the maximum (eotemporal) value and zero (the atemporal value) depending on the relative velocity of the observer.

One is free to think of laboratories drifting at velocities approaching that of light but nature is not so obliging. (Figure 1b). We do not have massive bodies that move at relativistic speeds within the galaxy.* As we construct our imaginary

*Leslie Marder provides reliable calculations on space travel. With an acceleration of constant magnitude g for one-half of the journey, then a constant magnitude g deceleration for the other half, a spaceship would reach 95% of light velocity mid-journey and get to the nearest star (4.2 light-years away) in 3.5 years of its own time and 5.9 years of stay-home time. If it were to continue at the same constant rate of acceleration, it could approach light velocity to 1 part in 10^{12} and get to the Andromeda galaxy (2×10^6 light-years away) in 28.2 years of spaceship time and two million years of stay-home time. But these are paper and pencil exercises.

For various reasons, chemically propelled rockets would not do, even if all but one part in a million of the initial mass of the spaceship would be fuel. Photon rockets and arbitrarily long periods of coasting would be more practical, especially if interstellar matter could be used as fuel. But acceleration provided by photon rockets has a very small magnitude, therefore many generations of astronauts would have to make up the migrating colony. Their trip to Andromeda might take millenia in their own time – at speeds that are a few percent of that of light.

THE STRUCTURED MINKOWSKI DIAGRAM

Motion in the X-Y plane of this diagram may be represented by curves (world-lines) in the X-Y-T space. Our initial reference is the here-now of an inertial human observer located at origin O of the coordinate system. At instant t he makes a rapid mental survey of his world, using his clock time as his time reference and plotting his units of time along the T axis.

Special relativity theory permits him to imagine any kind of motion in the X-Y plane as real, and hence to take any world-line as a possible one, as long as that line remains within the light cone. But nature does not permit him to select entirely freely such objects as would perform the motions represented by the arbitrary world-lines. This fact challenges the unqualified generality of the ordinary, unstructured Minkowski diagram as a representation of time and space.

World-lines in the light cone must be those of particles with zero rest mass. For reasons discussed in the text, such particles determine an atemporal Umwelt. The regions limited by, and external to, the light cone are also atemporal: they constitute the universe of cosmic, absolute unrest. In the vicinity of O they are contiguous with the atomic and the perceptual chronons. Were it not for the human observer, nothing in the world of photons or neutrinos could determine the kind of time which the T axis is normally assumed to represent.

World-lines that pass through the prototemporal regions close to the light cone must be those of elementary particles, they cannot represent the motions of trains or spaceships. These regions comprise the world of "imperfect speakers." In the vicinity of O the prototemporal regions are contiguous with the prototemporal present of the microstructure of time perception. Were it not for the conscious observer at O, the time axis T could represent only prototemporality.

A world-line which is mostly within the eotemporal cone is likely to be that of an astronomical object, but, in any case, of a massive body. This region comprises the Newtonian universe of local, relative rest. In the vicinity of O the eotemporal physical world is contiguous with the eotemporal present of time perception. In the absence of a living observer at O, the T axis could represent only pure succession.

World-lines of moving, living organisms are likely to lie within a narrow biotemporal cone. As we shall see in the next chapter, it is the necessary inner coordination of living matter that inserts a meaningful

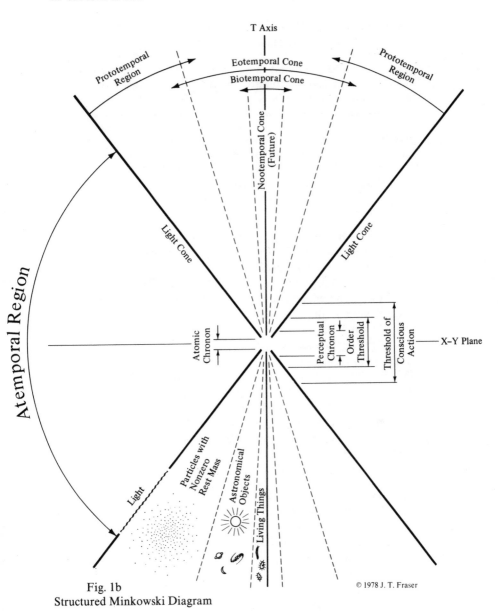

Fig. 1b
Structured Minkowski Diagram

© 1978 J. T. Fraser

now in the pure succession of the eotemporal world, and hence author-
izes the assignment of presentness to point O.

The world-line of the observer is the T axis itself: it is a nootemporal cone that collapsed into a line. Only the mature mind can provide
that sharp polarization of future, past, and present which is one of the
necessary initial conditions used in constructing the Minkowski diagram.

We have discovered that in this picture the observer is not an
ornament, placed there as a heuristic aid for people who cannot think in
abstract terms. Rather, the observer is the determinant of the world
representable by the diagram. It follows that the geometrical simplicity
of the unstructured Minkowski diagram is badly misleading. When the
hierarchical nature of time is neglected, it becomes immensely difficult,
if not impossible, to reconcile the meaning of t or τ of physical theory
with the significance of human time.

---

observers of the two reference events mentioned earlier, we should
select the first astronauts from among physicists; then they must
be changed to heated plasma or gas; then to rapidly traveling mu
mesons and finally to photons and neutrinos. The clocks used by
these disembodied experimenters must also change from eotemporal devices to prototemporal ones and finally to photons which,
alas, cannot measure time in their own Umwelts.

During this series of tests the observer regressed from an
eotemporal to a prototemporal and finally, to an atemporal one.
The need for the regression cannot be removed from the experiments. Were it otherwise, we could send a kitchen clock across
the earth's orbit and demonstrate on television, for all to see, that
upon its return the traveling clock is behind the stay-home one.
Performable experiments on time dilation remain tied to microphysical phenomena, or else they involve extremely small effects
in the behavior of extremely large bodies.

Let us assume now that we are ignorant of the notion of
time but know relativity theory, have a kitchen clock whose
settings we know how to read, and are ready to observe the two
reference events mentioned earlier.

We already know that under some conditions the separation between them is measured as zero. There is nothing particularly new about zero separation because the fundamental substra-

tum of the universe is atemporal. We also know that it is possible to measure the separation between those two events by any number of clock ticks from zero up to a maximum value. We might then postulate a new maximal principle of physics: massive matter makes clocks indicate a maximum number of ticks between events at rest. The influence of massive matter produced something new that did not exist in the prototemporal world. Using a phrase borrowed from spectroscopy, in the vicinity of matter the degeneracy of the atemporal world is removed and a scheme of well-defined maximal separations among events are revealed. This newly discovered capacity might even be given a name: the creation of eotemporality. Inertia, which prevents matter of finite restmass from becoming atemporal, will then not be associated with resistance to change in states of motion but recognized instead as a principle of evolution that directs nature away from the absolute motion and ceaseless unrest of the atemporal world.

In the eotemporal world space and time are clearly distinguishable by their experimental asymmetry, as may be demonstrated from special relativity theory. Consider that a traveling rod will undergo a relativistic contraction while on a rapid journey with respect to my inertial frame. A clock traveling with the rod will also undergo a relativistic change, it is time dilation. But upon their return the rod will be found at its original length whereas the clock will have fallen behind the stay-home clock. There is no way whereby, through further motion of any type, the clock that fell behind could be made to catch up with the stay-home clock. It seems, then, that aggregate matter which arrests the absolute motion of light also introduces a definite asymmetry between time and space: motion with respect to an inertial frame, at rest in the average matter distribution of the galaxy, leads to an absolute change in the measure of time but not in the measure of length.

Could a similar asymmetry be identified in the prototemporal world? The first, rather predictable difficulty there is to identify the parameters that would correspond to what length and time are in the eotemporal world. The best we could hope for are ill-

defined distances and ill-defined periods of time. Our measurements must be statistical and probabilistic. It is known that some
thermodynamic quantities, such as heat, are effected by relative
motion whereas intensive variables, such as temperature, tend not
to be. Though the details are far from agreed upon and the debate
is continuing, a reasonable expectation is that some quantities will
be identified as motion-dependent and some not so. Only those
quantities which undergo relativistic changes while in motion
could be used for a test of asymmetry between spatial and
temporal measures, and one would expect that when the issues are
clearly understood and resolved, the prototemporal world would
display an intractable ambiguity between space and time.

     Pure succession finds its formal expression in certain equa-
tions of physics which are invariant under time reversal. In fact,
all laws of physics, except the regularity known as the expansion
of the universe, and the second law of thermodynamics, imply
worlds of pure succession. We shall later see that arguments of
physical cosmology cannot supply reasons for the origins of an
arrow of time, and that the association of that arrow with the
second law of thermodynamics is also wrong.
     Time reversal invariance in physics has long been known.
Simply put, it states that physical processes are unresponsive to a
change in sign of the physicist's t. This fact has given rise to the
belief that the physical world somehow functions outside time.
The correct and defensible view is that time reversal invariance in
physics implies the pure succession of eotemporality, or even the
curious time of prototemporality – it does not imply timelessness.
The idea of timelessness has been used rather uncritically both
outside and inside physics. It has been claimed for microphysical
processes, for reversible macroscopic processes, for the description
of the microstructure of time perception and even for certain
forms of life.
     The mixing of time measurements appropriate only sep-
arately to proto- or eotemporal Umwelts, and in turn their mixing
with spatial measurements appropriate to these Umwelts, have
given rise to many curious conditions. For instance, in the

relativistic substratum of four-space, time and distance measures are incommensurable. Using the imaginary unit i to distinguish the temporal from the spatial metrics assures that, geometrically, the time and space coordinates are always mutually perpendicular (regardless of the number of spatial dimensions), hence time and space appear mutually point-like when mapped one into the other. But this neat mathematical sleight-of-hand only hides the rich physical reality that is beneath it all and through its Platonic simplicity, it easily leads to confusion.

To have some idea of just what happens, consider that it is the physics of light that makes possible the mutual transformation of temporal and spatial measures, and thereby gives meaning to the four-representation of the Lorentz Transformations. The quantity c usually bears the dimension of velocity, but physically it represents that process which makes the atemporal world one of absolute unrest. Our usual paradigm for time measurement is the eotemporal laboratory clock, and the symbol t of the Lorentz Transformations is usually meant to represent the readings of such a clock. But, depending on the physical Umwelt, the meaning of that t need not be eotemporal: it may also be prototemporal and, when zero, atemporal. We have seen that in the three physical Umwelts, the degree to which space and time are distinguishable is different. The asymmetry of traveling clocks and rods is eotemporal; in the prototemporal, space and time measures are only weakly distinguishable; in the atemporal, not at all. Thus, even within special relativity theory there is a hierarchy of distinctions between space and time. When embarking on a program to construct a unified formalism which would embody general relativity theory, quantum mechanics, and special relativity theory, we are seeking a scheme that would unite in one well defined and (presumably) conceptually simple form three different temporalities, and three different bifurcations of space and time. It is not surprising that quantum field theory is difficult.

Future and past make sense only if there exists a now with respect to which one can specify events to come or events that have passed. There is nothing in the physical Umwelts that would correspond to our ideas of nowness. This is another fact which has been taken to imply that the physical world is timeless (or, if one holds with physical reductionism, that time is an illusion). The correct and defensible interpretation is that eotemporality is a world of pure succession, with no direction of time and no nowness.*

In a world of pure succession one ought to be able to give meaning to a period of time with well-defined limits. After all, although $s = \frac{1}{2}gt^2$ is insensitive to the sign of t, it is surely not insensitive to its magnitude. But an ending or a beginning must again refer to a "now", even if that "now" coincides, for instance, with the beginning. Thus, whereas finite periods of time in the eotemporal must be identifiable, the boundaries of these periods cannot be called beginnings or endings, even though our language does not accommodate such a curious state of affairs.

We have no specific principles that would bridge the types of causations that characterize the proto- and eotemporal worlds. Causation in the prototemporal physical world is probabilistic. As we cross into the world of macroscopic matter causation become deterministic and single valued: one cause, one effect. ** We must assume that, because of the hierarchical structure of nature

---

*See the critique of "time asymmetry" on p. 24.

** The ancient game of dice may be used to show how probability as belief, working in the eotemporal world, changes into probability as the appropriate lawfulness in the prototemporal world.

The motion of the dice should surely be classed as that of a massive object, yet the game remains aleatory; the prediction that a numeral will turn up, on the average, 1 in 6, is a statement of belief. We might attempt to change this belief into certainty by improving our understanding of the dynamics of dice throwing. Let us weight the dice and balance it well; design a throwing machine with great care; operate the experiment in a vacuum and anchor the devices to solid rock. Eventually, for known initial conditions, our predictions will become fairly reliable, even deterministic, with a bit of statistical residue left over for experimental error. Reduce this error by making the dice completely balanced. Instead of the usual slight indentations use painted dots. Since one dot weighs less than 6 dots do, paint six of them on each side but use different color coding. Paints are of different densities so use radioactive labeling instead, but do calculate the reaction of the dice to the emission of particles. By the time the dice become ideally true, the faces will have become indistinguishable. We will have left the eotemporal world for the prototemporal one.

deterministic causation subsumes probabilistic causation as its immature form, just as eotemporality subsumes prototemporality. In the eotemporal world time still does not have a direction: therefore causes and effects are interchangeable. Again, our language does not have words to accommodate these facts.

The prototemporal and eotemporal physical worlds are connected through a number of principles. One of them is the famous uncertainty principle of Heisenberg. It asserts that increasing the accuracy of measurement of an observable quantity in quantum theory necessarily produces a decreasing certainty (increasing intrinsic error) in the experimentally obtainable values of other observables. For instance, energies or momenta of macroscopic bodies, or times and positions of macroscopic events are well defined in the eotemporal world; that is, they may be measured with great accuracy in terms of the absolute values of these quantities. The intrinsic fuzziness of the prototemporal conditions vanishes in the numerical vastness associated with the macroscopic world. When we attempt to use eotemporal language (the signs and symbols of macroscopic physics) for specifying the happenings of the prototemporal Umwelt we find it too rich, too exacting for the task.

It has often been pointed out that the microscopic world of elementary particles is characterized by reversible processes, whereas the aggregates of such particles show a one-wayness, or irreversibility in their collective behavior. In the same, unexamined terminology that I already criticized, the question is usually asked: how do the timeless conditions of microphysics change into the temporal conditions of macroscopic physics?

From what we learned earlier it should be clear that the question "how does the timelessness of microphysics change into the time of macrophysics" includes an erroneous assumption, namely, that the nootemporality (taken to be the world of time) is contiguous with the physical temporalities (taken to be the world of timelessness). But the issue does not involve a timeless/time interface at all, but only the transition of the time of cardinality to the time of ordinality, from the prototemporal to the eotemporal.

It is a change from a world of indistinguishable objects to one of distinguishable objects, and the step resides precisely in that change.

Many clever ideas have been put forth to explain the change in temporal behavior from microphysics to macrophysics. Foremost among them is the appeal to Boltzmann's famous H-theorem which states that, in the course of time, isolated ensembles evolve toward more probable, because more mixed-up, states. We shall find that the association of the trend implicit in that theory with the direction of experiential time is indefensible. The continued rearguing of statistical thermodynamics as the basis of the emergence of "time" from the "timeless" witnesses the malaise that these arguments leave in their wake. It should not be surprising, but rather encouraging, however, that no satisfactory theoretical reasons have ever been found in support of conditions to which nothing in nature corresponds.

4   Two Arrows of Time for the Price of One – Part I

We have learned earlier that future and past make no sense without a now. It is, therefore, rather curious that whereas the non-existence of a physically meaningful present is universally admitted in the natural philosophy of time, the physical basis of an arrow of time has been energetically sought. The reasoning in support of a physical arrow is based on the widely held belief that the monotonic average increase of entropy production in closed systems, as expressed in the second law of thermodynamics, represents a one-wayness which, in some yet unspecified way, underlies our experience expressed in the metaphor that time flows from past toward the future.

Under the heading "Two Arrows for the Price of One" (of which this is Part I) I intend to show that the thermodynamic argument is specious and that the highest temporal level of the physical world remains one of pure succession without a now *and* without an arrow. Both the presentness and the preferred direction develop only with the evolutionary emergence of life.

The argument for the thermodynamic arrow employs the rich notion of entropy, a concept which first arose in the context of the theory of heat; later, it was reformulated in statistical terms that permitted its employment for the description of probabilistic behavior of aggregates of elements of many kind. Subsequently, the idea was extended to include the quantitative theory of information, paving the way to the formulation of entropy principles that could be applied to self-organizing systems, learning, economic life of society, and possibly to the creative expressions of the mind. Through its impressive career, entropy remained a measure of disorganization of a system with respect to a more organized state of the same system. Specific formulations of entropy principles exist for such diverse uses as steam tables, transfer of messages along a radio link, learning behavior of rats, and the economics of commodity production. There is no single unit or even physical dimension for entropy that would be common to such different uses, and specific formulations of the second law must obey different principles depending on the system to which it is applied. That eventually appropriate common units will be found so as to permit some type of direct comparison among temporal processes in boilers, telephone lines, rats, and human economy is, at the moment, only a belief. It is a reasonable belief, however, for the principles that underlie the many uses of the second law appear to be common.

In one of its mature forms the second law of thermodynamics states that the entropy of a closed system increases in time in a statistical way. There may exist local and transient conditions within a closed system such as to lead to entropy decreases (or to no change) but, after a long enough period of time the entropy of the system will be found to have increased. By closed system is meant one so bounded that no energy or matter may be exchanged between it and its environment.

The power implicit in the generalization of the second law caught the attention of Eddington, the great English astronomer, who perceived in the trend implicit in that law the possible physical basis of our sense of time. He coined the phrase "the

arrow of time" to express the one-wayness implied by the law, a
direction of change which, he wrote, is "vividly recognized by our
consciousness."

The second law is not a statement about changes in any
specific system but only about changes with time of an abstract
parameter, entropy, which in its turn is presumed to be intelligible
for many actual systems. Thus, thermodynamic changes are under
(at least) dual control: the restraints of specific laws, and the
restraints of the second law. It is this fact of multiple control that
permits the application of the law of entropy increase, at least in
principle, to physical, biological, and perhaps mental processes as
well.

Closed systems must be regarded as smaller than the
universe, because they are, by definition, surrounded by an
environment with which they exchange neither matter nor energy.
The usual model is a box with a small world in it. For the purposes of this chapter, that small world will not contain living
matter. By the specifications of the second law and under the
restraints of the laws of physics and chemistry, the entropy of the
box is expected to increase from an assumed low entropy state to a
higher entropy state, from a highly organized one, to one of
thorough mixed-upness. It has been broadly claimed that this
direction of change coincides with what we feel to be the direction
of passing time.

It is fortunate that the box is smaller than the universe, for
this permits us to observe it from the outside and declare its
behavior to be consistent with the demands of the second law –
according to our sense of time. Meanwhile, inside the box, the
processes run their course under the control of rules which,
collectively, form the laws of physics and chemistry. How do the
dual controls function?

O Beginning with very general considerations (such as that no
force can be manifest or observable without moving some mass
some distance in some direction) and through the clever use of
dimensional analysis, C.F. Muses has shown that all closed
physical systems tend to change so that the variation in entropy is

minimal. While the system overcomes the effects of inertia, hysteresis, friction, or resistance of any kind, the entropy increase demanded by the second law obtains along paths which, compared with alternate virtual paths, are minimal. The box of our example, containing mechanical processes, will eventually reach equilibrium; entropy variation around that final state will then be stochastic and the entropy of the system will remain constant. An observer looking into the box will note complete disorganization.

○ Yourgrau and Raw concern themselves with chemical systems close to equilibrium. The rates of entropy production of stable chemical systems under such conditions are minimal. If there is any deviation from that stable state, thermodynamic controls will tend to reestablish stability through "theorems of moderation," such as the principle of LeChatelier.*

○ In physical chemistry, Ilyia Priogogine studied steady state conditions of systems far from equilibrium. He maintains that for such systems the conditions I mentioned above still hold, at least for certain restrictive conditions. By a theory of minimum entropy production, a universal criterion of thermodynamic stability is provided for stationary states, in general. Such states are immune to small perturbation and therefore, he maintains, they more readily represent real life problems than do unstable states.

The paths of these arguments meet in a single conclusion. The long-term fate of a closed system is expressed by the second law, but the final state of equilibrium is reached along paths of minimal rates of entropy increase. At the end those minimal rates will reach their lowest possible limits: at equilibrium no further entropy increases obtain.

The theory of minimal entropy production as a universal principle of steady state systems is still being debated. At our present level of understanding it strongly suggests that closed

---

*In a system in equilibrium, when one of the parameters which determine the equilibrium is made to vary, the system reacts in such a way as to oppose the variation and partially, to annul it. In electrodynamics Lenz' law is an exact analogue of LeChatelier's principle. And since indistinguishable members of any class determine a prototemporal Umwelt, we even have an analogous principle in economics, known as Volterra's theory of population.

systems function under the simultaneous controls of a single entropy increasing principle and a body of specific laws that tend to keep the rates of entropy increase minimal. Opposing the arrow demanded by the second law there is a class of arrows demanded by specific laws. The actual path of entropy increase is determined by the presence of two opposing trends. It should be immediately added, however, that the only reality is the development of the system. The opposing arrows are analytical tricks, as are the action and reaction forces of Newton's third law of motion.

Even brief reflection will reveal that it is not really possible to construct a box that would correspond to the ideal specifications of a closed system, except by removing it from our universe. We must, therefore, make the box part of our universe and take the isolation of its contents only as an approximation. But then, whatever happens in the box must include directly or indirectly our opinion about the universe, and the internal happenings cannot altogether be separated from our particular choice of cosmological models. Let us, therefore, enlarge the box until it becomes coextensive with the universe. We might make the reasonable assumption that the universe is thermodynamically closed, although it has been argued that it may be open with no upper limit for entropy. We shall see that neither assumption is really relevant.

In an unusually perceptive work, P. C. W. Davies sought to identify the sources of time asymmetry in the physical world. Here "time asymmetry" is used to signify a present-less time, as is our pure succession of eotemporality, but unlike eotemporality, this time has an arrow. Although we found that nothing in nature can correspond to such conditions let us assume, in the framework of Davies' inquiry, that it can.

"It is a remarkable fact," he writes, "that all the important aspects of time asymmetry encountered in the different major topics of physical science may be traced back to the creation or end of the universe." Since the direction of time cannot be attributed to any built-in asymmetry of physical law applied to local conditions, it must be attributed to the universe at large; and if it is, then it must be the consequence of peculiar initial and/or final conditions.

Davies also gives reasons why the universe is not at all distinguished by such peculiar initial (or final) conditions but is, instead, a typical member of an ensemble of universes. As homework, the reader is left with the task of interpreting what is to be meant by an ensemble of universes.

Curiously, they do make sense. One may take, for instance, Davies' own model which is really a two-step universe: one forward and one back, making for a closed-loop time. Or, one may take the corollary, the continually oscillating universe of expansion and collapse, expansion and collapse, in saecula seculorum, from generation to generation.

He notes that for such models, for each cycle "the asymmetry between the two ends of a cycle of expansion and contraction is completely arbitrary, and must be accepted as a fact of nature." If one tries to impose time symmetry on the model (so as to make it consistent with the arbitrariness of time direction) the constraints make the results incompatible with observational data. "However, if one is prepared to accept the ... exotic possibility of topologically closed time, then good agreement with the observed universe is obtained."

It cannot be our task to decide here matters of technical detail in physical cosmology, but some important remarks may, nevertheless, be made.

Topologically closed time is the idea that time re-enters itself upon the completion of a cosmic cycle and the end becomes a beginning – not a new beginning in linear time, but somehow the beginning which to us now appears to have been in the past. In terms of the linear experience of our human time this makes no sense. However, it does become a straightforward and clear issue once one assumes that no physical process, including the behavior of the physical universe at large, can determine Umwelts higher than the pure succession of the eotemporal. The topological closure of time, then, does *not* signify a loop where future become past upon the completion of a cycle, but is only a way of saying that in its highest Umwelt the physical universe is eotemporal. If judged entirely from within, its ending and beginning cannot be told apart. But if this be the case, as I have been arguing it is, then any attempt to tie the direction of nootemporality to an entropic description of the universe is doomed to failure. Perhaps, while talking about physical cosmology, we made the mistake of neglecting the cosmologist.

Not having found an arrow of time in the local or even in the cosmic closed box, let us backtrack to the local box and relax the requirements on its walls, so that information, matter, and energy may now pass through them. If the box contains some structures that, through some yet unspecified ways, are capable of self-organization the supply of energy and/or matter flowing into the box will make its continued self-organizing activities possible. An enclosure of the type described is called an open system. On the average, the entropy of a self-organizing structure (an open system) will decrease in a statistical way. It might increase occasionally, but if we wait long enough, we shall find its organization improved and, hence, its total entropy decreased. If an open system is placed within a larger, closed system, the entropy increase of the larger system, if observed long enough, will be found to have more than compensated for the entropy reduction by the open system within. Thus, the presence of an open system within a closed one will only urge the final thermodynamic demise of the closed system. Note, however, that "final" is to be understood in terms of the time sense of an external observer.

Open systems do occur in the inanimate physical world. It has been shown, for instance, that the growths of typhoons over the Pacific Ocean represent entropy decreasing physical processes, as do the growths of vertices in viscous fluids in general. Other examples include certain moving striations in the positive column of gas discharge, and the lining up of atoms in lasing action just before the emission of a pulse. However, entropy-reducing physical processes are rare, they are not systematic, and they are never self-sustaining for any appreciable length of time.

Much more impressive and very important for our purposes, is a class of closed systems which probably represent the highest level of self-organization to be found in the inanimate world. I am thinking of the growth of crystals, both solid and liquid, both inorganic and organic. There are reasons to believe that in crystals the capacity of matter to oppose the trend of disorganization in nature reaches its upper limits, hence its fiasco. Indeed, while physical processes do minimize the rates of entropy increase demanded by the second law of thermodynamics, they cannot, figuratively speaking, turn back that trend and consistently absorb entropy. For that capacity, we shall have to look to some new processes.

Attractive as the idea may be that our sense of time with its hallmarks of futurity, pastness, and presentness derives from the physical world, we must conclude that this cannot be the case. We have examined the levels of temporality consistent with physical law, and found that they lack temporal direction. In the inanimate world, "boundless and bare, the lone and level sands stretch far away."

IV   The Birth of Life, of Death, and of the "Now"

LIVING MATTER is controlled by the laws of the physical world, as are volcanoes or rolling stones; organisms are so true to their environment that they even model geophysical periodicities, such as that of the rising and setting sun. Yet there is something peculiar to life which, in some subtle and some coarse ways, is expressed by all things alive. The moon, for instance, is a dog's senior in age by ten orders of magnitude and more massive than any canine by some twenty-three orders of magnitude. Yet, when the dog bays at the full moon with fury and fear, its motion and its attitude express a restlessness and striving which makes that inconsequential mass of fur and noise substantially superior in sophistication and autonomy to that pale-faced heavenly body. How did such a state of affairs come about?

In this chapter I shall deal with the problem of the origins and subsequent evolution of life in terms of the theory of time as conflict.

1   From Inanimate to Living Matter

There seems to be little doubt that living forms came into being from non-living matter, perhaps three billion years ago. But what exactly we mean by biogenesis (the emergence, or birth of life) is not at all clear. Creating new human life from food, air, sunshine, and the good works of two people is a popular enterprise, yet this is not what we mean. One could perhaps construct the zygote of a garter snake from chemical elements. This would also amount to the creation of new life and, technically, quite a remarkable achievement; yet in principle it would be nothing more than copying what nature had already created. Perhaps someone could make a serpent, unknown on earth, one that breathed ammonia vapor and reproduced by binary fission. This would be an example of what Joshua Lederberg called "algeny," short for genetic alchemy, which is a remodeling of existing genes. Before the creator of the new serpent could get credit for his work he would have to convince his peers that what he had made was truly a new form of life and not simply a moribund mass of cells. It seems, then, that by biogenesis we mean the coming about of organisms that behave according to the laws of biology as they are understood today. Should our understanding of life change, so would our ideas of what constitutes the birth of life. Where and how we look for the origins of life depends very much on what we judge life to be.

There are currently many ingenious ideas about the possible origins of life. Almost all of them concern the ways in which the chemistry of self-reproduction might have come about, and that is certainly an important detail. However, if the minimal possible self reproducing unit corresponds to the schematic organization suggested by John von Neumann (chassi, box of instructions, machinery to make another chassi, instruction replicator, and clock) then self-reproduction is already too complex a process to have arisen by what must surely have been a chance encounter of conditions. Also, there are no reasons why the chemistry of primordial life must be assumed to have been the same as even the most primitive life known today.

It is the thesis of this section that our most distant ancestors were molecular clocks whose frequencies fitted the spectral niches of their environmental periodicities. They drew from the environment the energy needed for their continued oscillations and assured thereby the partial autonomy of primordial life. I regard this autonomy more fundamental than, and developmentally much prior to, the need for reproduction. The question is, then, how did the capacity for sustained oscillations arise in the physical world, where open systems are only transient occurrences and never stable functions.

In the following we must share the genetical view of life which maintains that the know-how of living resides in the genes, and that the soma is never more than yet another test for the viability of genes, through which they reproduce and spread. The phenotype is but a hefty interface between the genes and the larger environment, with the body being that part of the gene's environment wherein the interaction between them and the environment is most evident.* It is thus the ancestry of the gene that one seeks. We search for the origins of that stunningly stable system which is shared by all living organisms, underlies the identical metabolic machinery of all life, and guarantees the identical methods through which all organisms handle inherited variations. It is well said that whereas the soma reproduces, the gene replicates. And if one looks for structures that replicate reliably and stably, one can hardly find better candidates than crystals.

Joseph Needham writing on the integrative levels of nature (1944) spoke of "mesoforms" which occur between successive levels of organization; "thus between living and non-living matter the realm of the crystalline represents the highest degree of organization of which non-living matter is capable." J.D. Bernal spoke of "generalized crystallography as the key to molecular biology." That crystals were in fact our distant ancestors is the thesis of Graham Cairns-Smith's inquiry into the origins of life. "The primitive genes were not just like crystals," he writes, "but actually were crystals." But the chemistry of crystalline matter is quite different from the biochemistry of genes; hence, he postulates that "at the core of primitive organisms there was some kind of solid state mechanism that stood in for the complete DNA-RNA-protein system of modern organism." He then suggests that the inorganic constituents of crystals were replaced by the nucleic acids and proteins that form the basis of our present biochemistry, in a process of "genetic metamorphosis." What early genes retained from crystals was structural organization.

---

*Thus, the water and the heating systems of a house with a single person living in it are also part of the environment of his genes, even though that relationship is not immediately evident. The soma does not have well defined limits because it influences, and is influenced by, processes in regions well beyond the confines of the visible body, such as, for instance, man's scratching the surface of the planet Mars.

During the Precambrian epoch when life is believed to have come about, crystalline substances were surely in abundance, washed, one would presume, by the waves of the primordial broth. And that broth itself was undergoing chemical evolution, producing certain molecular preferences. Remembering the speculations of Bernal and how microcrystals are likely to have absorbed the molecules of that soup, the bold suggestion by Cairns-Smith of a takeover of crystalline structures by organic substances does not sound implausible. Of course, the takeover would have depended on chance events, but so did so many other developments in the history of the earth. And if it did happen, there and then evolution left a crystal-controlled biochemistry behind in favor of a biochemistry controlled by organic compounds in solid, or liquid crystalline forms.

Inorganic crystals not only replicate stably but also oscillate reliably. Their modes of oscillation are numerous and cover a broad spectrum of frequencies from the electromagnetic domain to the audio region. Also, and importantly, there are numerous mechanisms whereby energy absorbed in one mode and at one frequency may be converted to oscillations in a different mode and at a different frequency. Few of our environmental periodicities today are rapid enough to approximate the natural frequencies of inorganic crystals, but our milieu is substantially different from that of the Cryptozoic earth.

During the Precambrian era the heaving, moving, and shaking of the earth was the rule. Overwhelming as it is in terms of human experience, we may easily permit ourselves to think of boulders of all sizes that would vibrate for five thousand years at a time; in relation to the age of life on earth this is only an instant, one part in a million. The noises that cheap science fiction motion pictures employ to convey the mood of the early earth may not be far from what the actual situation might have been, except for their meager scale of human decibels. Let us then assume that some of the crystals of the Cryptozoic Eon were able to absorb energy in one form or other and at certain frequencies, and to be driven to oscillate at other frequencies and possibly in different modes.

I imagine somewhere in the small hours of the earth, at that time only one billion years old, the coming about of molecular clocks. They needed to do no more than model with crude accuracy some of the cyclic variations of their environments, such as, perhaps, minute and rapid physical or meterological changes that accompanied locally the extensive and slower cyclicities of the earth.

Crystals, supplied with energy at some frequency or frequencies and in some modes of oscillation, positioned, for instance, in crevices whose sizes (so I would fancy) varied periodically, could remain in their positions longer if they were able to match the environmental changes. Since the regularities of these primordial clocks could hardly have fitted the temporal niches of the environmental spectrum exactly, there existed an error signal through which the environment could exert selection pressure on the clocks in favor of such configurations as were able to provide better adaptation to the cyclic changes of their surroundings. With the possibility of preferring some frequencies to others, natural selection supplanted accident. Current research holds analogous views on the origins of circadian rhythms: it is believed that the daily rhythm was not generated but selected by geophysical cycles, by exploiting already existent oscillating mechanisms, making the frequencies more precise, and conserving them. In the primordial mismatch between the expected and the encountered we might recognize the origins of existential stress that has characterized life through its known career.

The genetic metamorphosis envisaged by Cairns-Smith is likely to have had its dynamic corollary. Thus, the replacement of inorganic molecules by large organic ones made slow chemical processes available from which evolution could select. The way was open for the development of biological clocks with long periodic functions, such as those which we find in organisms alive today: they would meet the rhythms of daily temperature, light, and salinity variations, and the like. Daily rhythms in particular are ubiquitous; they are believed to be present in all living organisms, and permeate them to their basic structures. There is considerable evidence that, in the words of C.F. Ehret, the ubiquity of the daily rhythm is "defined at the level of gene action by a mandatory circadian cycle."

Curiously, it is not at all necessary to assume that the earliest forms of life had reproductive capacities. Reasons have been given by Alexander Oparin and others why living functions ought to be separated from reproductive ones, which probably did not evolve until the appearance of nucleic acid. The living functions of our stable crystalline ancestors might have been comprised mainly of continued oscillations appropriate to external cycles, replenishment of the energy needed to maintain these periodic variations, and crystal growth. To them we could apply what J.B.S. Haldane has said about bacterophages: they were "on the road to life but it is perhaps an exaggeration to call them fully alive."

That in our own epoch we cannot identify spontaneous processes that would correspond to the creation of life, may be due to the transient nature of primitive forms. Early living crystals were either reabsorbed by the amorphous organic and inorganic compounds that populated our earth three billion years ago, or else evolved into more complex and hence, presumably, more stable systems. As their legacy, they left with us the stable morphology of the DNA.

2 Extending the Spectrum of the Cyclic Order

Let us assume that semi-autonomous oscillating systems did come about in the Precambrian Era by some such means as has been suggested. Whenever the environmental conditions called for inappropriate responses from these early clocks, selective forces became operative and a natural clockmaking enterprise began.

The making of living timepieces, as here envisaged, has had built into it the necessity for complexification. The situation is somewhat analogous to the old wisdom that the more we learn, the more unanswered questions we are going to have. Consider that each improved imaging of an external periodicity did indeed

amount to a step in improved adaptation, because it constituted a
better fit into the cyclic spectrum of the environment. However,
each improvement necessarily increased the regions of interaction
between the organism and its environment. For example, with no
endogenous tidal rhythms the secular variations of the tides must
remain unrecognized by the organism. However, after the organism does develop an approximate adaptive cycle, it creates the
conditions for the workings of a new selection pressure that will
tend to correct the endogenous rhythm. Putting it differently,
there is now a new region wherein the expected and the encountered may, and are likely to, differ. In terms of the theory of time
as conflict, there ensues a necessary increase in existential tension,
and an increased need for adaptive refinement. In current evolutionary biology such a process is sometimes described as fine
tuning.

The adding of a new component to the harmonic analysis of
the environment also constitutes a new demand for improved
internal coordination. The coexistence within a single system,
comprising even a few clocks of different frequencies, makes
internal coordination necessary, if the various rhythms are not to
interfere destructively. This calls for increased internal control.
Inevitably, the increasing number of clock rhythms will produce
new periodicities, whether by linear superposition or by non-
linear generation, to which nothing in the external world needs to
correspond. The new, and entirely internal, cycles must also be
integrated with all others if, as before, the many cyclic functions
are not to be mutually destructive.

Thus, I would imagine evolving a continuously richer inner
landscape of rhythms. It would comprise a map of external
periodicities, a store of periodicities which have no external
correspondences, and a system of control to keep the clocks in the
shop from becoming mutually destructive. Since internal cycles do
often become externally manifest, the spectrum of biological
clocks upon which natural selection could work was bound to
become wider.

Indeed, the spectrum of biological rhythms is very broad.
Our skins respond to ultraviolet rays that tick 10^{16} times per

second; most organisms can pick up heat at 10^4 Hertz; the period of neural signals is between 3 and 10 seconds. Probably all living things on earth show circadian periods, including population clocks, organismic clocks, organs and organ cultures, tissue cultures, bacteria and, as mentioned earlier, genes. Lunar periods are also abundant in man and animal. Numerous organisms show circannual rhythms, while certain bamboos flower every 7 or 8 years. Many genetic rhythms are known with periods longer than the life of any individual member of the species. In many species breeding seasons are periodic and are carefully synchronized among members of the species. The awe-inspiring aspects of periodic group activities, such as the hunt, the migration of the group, or the rituals of mating are well known.

An aside on intellectual history is appropriate at this place. The ubiquity of circadian, lunar, and seasonal rhythms is to be expected. Throughout the career of organic evolution the lives of plants, animals, and men have been subjected to an unceasing and regular variation of light and darkness, heat and cold, the cyclicity of the tides, of the seasons, and of all periodicities of the earth. It would be incredible if the experience of three billion years of daily routine had not left its profound and unmistakable imprint on life as we know it today. That there are meaningful connections between natural cycles and the periodicities of life was accepted without question by civilizations which regarded man as part of the natural order and the harmony between internal and external rhythms has been a source of inspiration to sensitive writers through the ages. The idea that connections between natural cycles and the periodicities in human functions are at all significant was rejected, however, with the Christian emphasis on man as a being apart from nature. It is only during the last few decades that the issue of biological cyclicity was pulled out of metaphysical oblivion; science rediscovered the obvious and began to supply details previously unknown and unsuspected.

Our arguments on biogenesis and on the ubiquity of the

cyclic order, when taken together, suggest that life be described as a clockmaking enterprise. However, unlike clocks designated to assist man, clocks created by organic evolution do not serve but comprise living matter. One immediate methodological consequence of this view should be mentioned.

Much energy and ingenuity has been expended during the last few decades on attempts to identify biological clocks or, perhaps even *the* universal biological clock. It is only quite recently that some doubts have appeared as to the soundness of assuming that organisms possess single master clocks or even several discrete pacemakers. Both the theoretical and experimental work has been biased by emphasis on space. I believe that if we wish to study the temporal organization of living forms, they ought to be "timesected" rather than dissected.

Workers concerned with the physiology of biorhythm often use the metaphor of removing the hands of the clock so that they might find the clock. Reflection will reveal, however, that at each level of complexity the hands of a clock are indistinguishable from its clockness. The proper way to proceed is to pursue the levels of temporal organization and descend along the temporalities as far as one wishes.

One of the many ways one can timesect organisms is to observe the modes whereby they solve adaptive problems in relation to time. Organisms alive today do this through an immensely complex multiple-level tracking system which extends from biochemical, instinctive, and reflex actions of the individual to learning, and to species replacement. As in the (speculatively reconstructed) process of biogenesis and subsequent organic evolution, so in the life of species, we can observe increasingly finer adjustments made so as to fill the available niches in the frequency spectrum of the environment. L. B. Slobodkin and A. Rappaport see individual organisms and populations responding to external perturbations by a set of mechanisms which differ in their activation times. At the instant of perturbation a series of responses is initiated; when the rapid and low-stake mechanism fails to deal with the perturbation, slower response and higher stake mechanisms are brought into play. The slowest of all such responses, and the ones with the highest stakes, are modifications in the genetic endowment of the species.

As a class of phenomena, the broad spectrum of periodicities that extends from light frequencies to rates of genetic change, and which evolved (so I speculated) from the humble beginning of crystalline clocks, constitutes the cyclic order of life.

3   The Necessary "Now" of Life

A system which comprised coupled oscillators could endure only if the chemical and physical processes of the individual clocks were mutually non-destructive. Thus, as already stressed, the complexification of biological clocks had to be accompanied by the evolution of inner controls that assured their necessary coherence. When we say that evolution by natural selection created semi-autonomous aggregates of biological clocks, we mean that it selected from among confederations of molecules some such systems as were internally stable. The capacity of biological clock˚ systems to survive, that is, to remain alive, may thus be seen as categorically tied to meeting and maintaining certain conditions of coincidence: some events must be, and some others should not be, simultaneous.

Simultaneities may, of course, be defined for non-living systems; I have already spoken about such conditions at length. But in the physical world simultaneities can have no self-referential significance. It is true that certain chemical and physical changes can take place if, and only if, the simultaneous presence of necessary substances and conditions is assured; but should such coincidences fail to come about, no organizational autonomy is lost because there is none to lose.

In sharp and significant contrast, even the most primitive living forms must maintain certain simultaneities among their constituting processes, instant by instant. Such simultaneities need not be, and usually are not very sharp, certainly not compared, for instance, with the atomic chronons. They are not determined by light velocities, and they do not involve distances comparable to elementary particles; they are determined, instead, by transport and reaction speeds available within living systems. The necessity to secure and maintain simultaneities within living organisms is, nevertheless, a basic criterion of viability. This same necessity is also the determinant of a new temporal Umwelt, one with a "now." It is thus that the necessary inner coordination of living matter inserts a meaningful present in the pure succession of the eotemporal world.

4   Reproduction, Aging, Death

As I have already mentioned, there are some good reasons to believe that reproduction is not coeval with life but is a later evolutionary development. Consider next the arguments of John von Neumann that there is a certain level of complexity which a machine must reach before it can produce others of its own kind, of equal or greater complexity. Below that threshold it can produce only automatons of lesser complexity. Weaving these two separate threads into a stronger single thread, one would expect reproductive capacities to have evolved only after early life crossed a certain threshold of complexity. Before that threshold was reached our distant progenitors oscillated, crystallized, and slowly complexified and occasionally, no doubt, broke into pieces. After a critical level of complexity was reached reproduction became advantageous because, on the average, the offspring could become as well as or better adaptable than was its parent. Henceforth, organisms that could reproduce as semi-autonomous units were in a better position to survive than systems which simply grew and, accidentally, split.

The great advantages of reproduction reside in the increased variety of individuals upon which natural selection may operate. As we imagine generations of organisms selected for increasingly finer mapping of external periodicities, the spectrum of the cyclic order of life could expand only in two directions: toward the higher and toward the lower frequencies. I have already postulated that such an extension of the spectrum did take place. Necessarily, one would expect that practical limits at either end of the spectrum would sooner or later be reached.

Consider the human eye: it responds to oscillations of about 10^{15} Hertz, and represents the upper limit of adaptation of the cyclic order of life. It happens to be a response to electromagnetic oscillations. At the other extreme we find circannual clocks, and even clocks whose periods are several years; they respond to geophysical cycles. The workings of biological clocks are very poorly understood, but it is obvious that the morphology of clocks that oscillate at 10^{15} Hertz, or once or twice a second, or else with a period of 28 days, or of a year, can bear very little resemblance to one another or to the morphology of that ubiquitous clock, the circadian. It is thus that the widely different physical and chemical processes which are involved in making clocks of different frequencies force upon living matter a necessary division of labor, and its corollary, the development of different forms.

We learned earlier that the widening of the spectrum also calls for increasingly refined controls so as to keep the organism from becoming self-destructive. One must assume that the internal control system itself has certain limitations. Hence the evolution of closely coupled biological clocks is bound to reach an equilibrium through attrition by specialization. A boundary so reached is a dead-end in the development of the purely cyclic order of life.

The suggestion comes to mind that it is because of such limitations that the cyclic order of life had to give rise to the aging order. What evolved may be described as a devilishly (or divinely) clever practice which involves a "throw-away soma" and an "immortal germ cell." The technique calls for systematic regressions from the soma to the gene, that is, from phenotype to genotype, once every life cycle. Henceforth, it was the life cycle which became the basic unit of evolution, with successive life cycles connected through a single cell. These single cells carry the genetic instructions plus a sampling of the environment. Complex living forms thus periodically return to the ancient forms of life from which they arose, though they never reach back as far as the "bare genes" when nucleic acid replicated nucleic acid. Putting it differently, the genotype itself evolves. Throughout the career of organic evolution the life cycle became increasingly elaborate but the basic scheme in which the phenotype periodically vanishes and life reaches down to its ancestral form, was retained.

During the evolutionary division of labor, procreation came to be relegated to a specialized part of the body, sometime during the Precambrian Era. The major function of this specialized region has been that of passing genetic information unchanged, or almost unchanged, from generation to generation. Among complex organisms that stage of early life which corresponds to the fertilized egg, or to the asexual spore, was maintained through the eons with remarkable stability, while the rest of the body evolved to respond to linear change and to unpredictable stimuli. That is, the soma evolved to make rapid adaptation possible. In contrast to the soma, mammalian germ cells may be regarded as though they were immortal: they change very slowly and, upon their division, no dead bodies of the parents remain.

Molecules, as molecules, are not expected to age: hydrochloric acid is never young nor is it ever old. Its Umwelt is without a present. On first approach, we should say the same of the molecules of deoxyribonucleic acid, or DNA. But biomolecules are dimensionally immense. The chromosomal complex of a small virus, the bacteriophage T4, has some 200,000 nucleotide pairs; those of acquatic animals between 10^{11} and 10^{12} pairs, that of man and all mammals about 5×10^9 pairs. Biomolecules can reach molecular weights of 10^8 avd perhaps 10^{11} and, unlike small inorganic molecules, they are never closed. In terms of the necessity of internal coordination, they surely define a "now" and hence, determine a biotemporal Umwelt, even though rather weakly. Perhaps they represent the very sophisticated and stable progeny of the metastable structures (to be discussed later) that once populated the eotemporal-biotemporal interface.

What we note here is that the link that connects two life cycles is close to functioning in a world of pure succession. Indeed, much of primitive life, such as viruses and most bacteria, determine eotemporal conditions: they die only through chemical or physical causes but they do not die by aging. Thus, while the germ cell remains for all practical purposes unaging, the soma, which I described earlier as yet another experiment by the gene, ages and dies. It is in the soma that linear time evolved and came to inform the existence of all advanced organisms.

The emergence of the aging order of life signifies the fiasco of the cyclic order, as a means for improving the adaptive features of organisms by periodical programming alone. This limitation is a corollary of the incapacity of the cyclic order of life to resolve the steadily increasing conflict between the expected and the encountered, brought about partly by the very success of the evolving cyclic order itself. But again, there is a price to be paid for the new improvement. There is a further increase in the existential tension between growth and decay because aging organisms have more ways to become frustrated than purely cyclic ones.

Of the several reproductive methods the sexual one has been most successful, and among sexual modes the heterosexual one, possibly because it is capable of spreading new mutations among members of a species faster than such variations would spread by asexual means.* The evolutionary origins of sexual reproduction are obscure, but surely it is the result of the evolutionary division of labor and possibly coeval with the separation of the soma from the germ cell. Sexual reproduction and death by aging thus appear to have common origins. Both are fatefully inscribed upon the biological information passed on in the advanced forms of life from generation to generation.

In matters of such profound significance as sexual reproduction and death by aging, the racial memory is very active. For

---

*Some animal species succeeded in combining both methods. For instance, freshwater hydras, aphids, and some other animals reproduce asexually when conditions are favorable for rapid population growth with little genetic variation. But as the environment deteriorates, such as when the shortening period of daylight presages the coming of short winter days, the animal shifts to sexual reproduction.

instance, in seeking the origins of sexual reproduction Plato tells us in his *Symposium* that Zeus had sliced the ancestral four-legged creature in half as one would "slice an egg with a hair." Since that time the two halves that made up the original person (called equally well a man or woman) have been seeking each other. Whenever they meet the original unity prevails and new life is conceived. Somehow, the blue sky of Hellas did not suggest to Plato any link between mortality and sex.

Not so for St. Paul who lived under the same skies, and for whom "the wages of sin is death," an idea that came to be incorporated in the Doctrine of First Intercourse, better known as the Doctrine of Original Sin. In its theological form it echoed down through the centuries, sometimes implicitly underlying most ethical instructions, sometimes explicitly such as in Milton's *Paradise Lost* where, in the opinion of the poet Frederick Turner it is charted as the central event in the psychological development of man.

Some nineteen centuries after St. Paul, Freud perceived in the mind of man an ancient guilt, a profound conflict between, on the one hand, the death-ignorant unconscious and, on the other hand, the conscious awareness of the inevitability of death. One can only wonder whether Plato, St. Paul, and Freud but articulated perhaps a billion years ex post facto that ingenious invention of organic evolution: the development of sexual reproduction and death by aging, as complimentary aspects of a new method of adaptation.

5   Two Arrows for the Price of One – Part II

Part I of our discussion under this heading concluded that the highest temporality of the physical world is the eotemporal, one of pure succession. In this chapter we have extended our knowledge of the development of time by discovering that nowness is inserted in the pure succession of the physical world by the necessary inner coordination of living matter. We also learned in Part I that physical and chemical processes, at least when they represent steady state conditions, are controlled by two sets of opposing instructions. A single, powerful principle provides for the systematic increase of entropy; a large family of specific principles provides for keeping that increase locally minimal. In this section we shall watch the two opposing trends open up, as it were, into the unresolvable conflicts of growth and decay of life. I shall also argue that since keeping alive is important for living organisms, it is now possible to give one of the two opposing arrows a definite preference over the other one.

The systems considered in Part I were mostly closed systems. Open, or self-organizing systems decrease their entropy content with the same necessary consistency as closed physical systems increase theirs. The trend is statistical: open systems might possess local and transient conditions that lead to entropy increase, but after a period of time the entropy of the open system will be found to have decreased.

For an open system to be able to function, it must be included in a larger system which is sufficiently rich to supply its needs. To keep the horse alive it must be included in a system with a pasture, the pasture with the earth, the earth with the sun; a closed box may then go around the solar system, at least as a first approximation. As already stressed in Part I, for the larger box, when observed from outside, the prior rule of increasing entropy once again holds. If the box contains life, this will tend to cause the entropy of the whole system to increase faster than it would otherwise, overcompensating by the quantity of total degraded energy the quantity of upgrading effected by life.

Unlike in the rare examples of physical open systems, operating in a decreasing entropy mode is a necessity for living matter. If an organism ceases to oppose for an appreciable length of time the thermodynamic trend of decay, it also ceases to live. Though the systematic entropy decreasing modes of living matter do constitute insignificant local deviations from the general law of entropy increase, they are surely not insignificant from the point of view of an organism. For life, the opposition to decay is not a deviation from the rule and a blemish on the laws of physics, but a permanent and essential part and parcel of existence.

It follows that if we take as our reference the living system, then our sense of time parallels the decreasing entropy trend of life, that is, the growth arrow. If we take as our reference the inanimate world, then our sense of time parallels the decay arrow. But then it follows that either direction is arbitrary and only the simultaneous opposition of the two arrows may be anything truly important. And, most emphatically, taking our reference in life is scientifically as sound as taking it in non-living nature. It must also be remembered that it is the living self which formulates theories of time, and it is the living animal that ages and dies. Although the universe is immensely larger than a wolf, in the study of time the wolf that roams the cold night must be taken as completely equivalent to the universe.

Let us return to the inanimate mini-world of the closed box that we constructed in Part I and listen to the box-man in it. He is an eotemporal and hence a non-living physicist, who will report, tenselessly, that whenever something does happen he notices two sets of opposing rules: a universal one and a local one. Whenever changes take place both rules are operative, in analogy to Newton's third law of motion that specifies that whenever there is a force, there is an action and reaction which are equal and opposite; no action-reaction, no force. He will describe the second law together with the specific laws as cooperating to produce stability and to provide for the prevalence of steady state conditions and equilibria. As the net amount of useless energy increases he will watch his world degrade from an eotemporal to a proto-, to an atemporal one, as his eotemporal sentences running both ways, tenselessly, become incoherent, and finally, vanish.

Arguments not directly involving the second law of thermodynamics have also been introduced in the continued effort to find a physical basis for our sense of time. The most interesting one is that based on central wave propagation. One starts by pointing out that wave motion originating at a point always spreads outward; we never observe waves converging spontaneously to a central point. A piece of wood dropped in a lake will produce outward moving waves only; we never observe the waves converging in the center, throwing up the wood that happens to be

Happy Birthday. Courtesy, Recycled Paper Products, Inc., Chicago, Illinois.

IF A YOUNG DRAGON or any other living thing, ceases to oppose the thermodynamic trend of decay, it also ceases to live. This fact is an insignificant local deviation from the general law of entropy increase permitted by the statistical nature of that law. But it is not at all insignificant from the point of view of organism. The opposition to decay is a necessary feature of life. Shakespeare had his own understanding: "Love's fire heats water, water cools not love."

floating there. However, I may drop a large ring on a quiet lake and watch the waves that propagate inward, and demonstrate that the functions of mind and body can oppose the physical trends in nature. Truly, there are no spontaneous processes of this type, but life is not spontaneous in the physical sense. When plants drive their metabolism through the use of sunlight, as the sun impinges upon them from east, south, and west, and with that energy they grow flowers and watermelons, they perform a feat analogous to my dropping a large ring on a lake. From the point of view of life, time flows in the direction of peripheral wave propagation toward a center. Which direction we pick as that of our sense of time is as arbitrary a judgement as that in the thermodynamic case.

The belief that our sense of time parallels some kind of physical arrow is very deeply seated; that it derives, somehow, from the increasing entropy trend demanded by the second law of thermodynamics, has never been seriously challenged. Yet it is a stance (together with the cosmic and the central-wave arguments) which is patently wrong. The clearing up of this universally held but quite erroneous view is so important that we must make a detour to a side issue: how could such a misperception have such deep roots among so many keen thinkers?

Progress is said to be made in scientific interpretation if one succeeds in formulating an understanding of increased universal significance, in terms of which the prior theories will appear only as particular cases that hold true for limited points of view. This is what the theory of time as conflict does when it reveals how limited and arbitrary, hence untenable is the view which identifies our past-present-future direction with that of the increasing entropy arrow. Yet, there are many good historical reasons why this belief has been so universally held, just as there were excellent reasons for believing in an earth-centered universe.

(1) Our acquaintance with closed systems preceded by a century our familiarity with open systems. Scientific understanding tends to be based on historical precedence of knowledge.*

---

*Electric current is said to flow in a direction opposite to that of the electron flow because the concept of current was invented before we knew of the travel of

(2) We are submerged in an immense, inanimate universe. Scientific judgment has been informed mainly by the advance of mathematized physics, and is inspired by the facts of the physical world more than by the facts of life. It is true that the biomass of this earth is totally insignificant in comparison with the mass of the universe. But for hydrogen gas or for the moon a now is undefinable, aging is not a very emotional issue, and death is meaningless. The record of the moon's past "written in its pitted surface" signifies a past in terms of our sense of time, and not in terms of the eotemporal Umwelt of the moon.

(3) Our civilization is heavily mechanistic and our clocks are exclusively inanimate. Were our timekeepers organic, were we to keep time by

> Hot lavendar, mints, savory, marjoram;
> The marigold that goes to bed with the sun,
> And with him rises weeping ...

rather than by clocks which tick and tock, we would surely favor the opinion that time's arrow points along the direction in which the entropy of living systems change. Learned papers would note, however, that for some reason the vast universe, with its cold and lifeless stars, runs the other way. Some might even conclude that only the opposing coexistence of the two trends can be important for the study of time.

The evolutionary decomplexification of matter, the decay demanded by the second law of thermodynamics, derives from the dynamics of non-living matter. Likewise, the evolutionary complexification of life derives from the dynamics of living matter. But whereas inanimate matter cannot define a now, living organisms can and do. It thus becomes possible to have a prefer-

---

electrons. And, by chance, the direction of current flow was picked such that, as later discovered, it opposed the motion of electrons. Had the electron flow been discovered first, the direction of current flow would surely have been selected to correspond to that. External to a dry cell, electrons would still migrate from electrolyte to carbon, but the electrolyte would be the positive pole, the carbon the negative pole.

ence; it is in terms of the self-interest and integrity of the organism. Life thereby acquires or, more accurately, creates the directionality of time. The entropy minimizing and increasing principles of the physical world may be seen as the roots of organic growth and decay, but no arrow of time can become explicit except through the autonomy of life.

Just as the integrity of individual organisms defines a meaningful now in the eotemporal matrix of pure succession, so the totality of life on earth defines a collective present. Information travels by means of chemical and physical signals whose propagation rates determine the degree of coherence among the actions of organisms, and whose comments carry the messages of life and death. When so cemented, all life on earth defines a biosphere with its peculiar systems properties, including the opposing arrows of growth and decay.

Let us now think of an organism as an autonomous vehicle of signs and signals which, upon certain kinds of excitations and after a lapse of time, will take some action. For certain macromolecular forms of life the delay may be only a few milliseconds; for simple metazoa it may be minutes or hours, for man it may be years, for society centuries; genetic responses may take millenia. Such delayed connections are identical in their manifestations to those of final causation; the purpose of the nest built today is to provide a place for the young birds in the future. Thus, with the emergence of life, the deterministic causation of the eotemporal world gave rise to the final causation of the biotemporal world.

We may now also identify the physical roots of futurity and pastness. Let us recall from Part I and Figure 1b that the future of a here-now contains events which are unknowable because their origins are in the atemporal absolute elsewhere. The same argument does not hold for the past, and hence we have an asymmetry. But this asymmetry is only potential because the highest Umwelt of the physical world is one of pure succession, without a present to which the asymmetry could be referred. As the contents of those atemporal region evolve, the polarization of future and past becomes stronger, or so one might think, because the store of possible future events has increased. But this increased richness of the future is only potential since we still do not have a present. The physical asymmetry becomes actual after the emergence of life, but remains of very little significance – except for people who are radio controllers for missions to distant planets.

Instead of a very slow and gradual evolution of the physical universe, consider the evolutionary history of sense modalities. The oldest ones are touch and taste; the oldest means of signaling are by pressure, heat, and by chemical substance.* Hearing is probably the next oldest sense, with hearing organs distributed at various positions on the body: on the forelegs (as in katydids and crickets), on the posterior of the body (as in water boatmen) and on the abdomen (as in grasshoppers). In vertebrates, however, the ear came to evolve in association with navigational control and was packaged in the head; its sensitivity greatly increased and became differentiated by species. The type of signals and organism is able to receive determine the amount of future time it has to react to the presence of friend, foe, or food. Signals by touch make this future perspective very brief, hearing extends it. In general, then, as organic evolution created the multiplicity of living forms, the separation of future from past became better defined.

Perhaps 500 million years ago, at a stage represented today by worms, our ancestors formed light sensitive regions in their epiderms. Since then, light has served as the foremost link between biological clocks and the Zeitgebers (synchronizers), as well as among organisms. Light sensitive spots are surely of best use in those segments of the animal which encounters the unknown world. Like Zeno's arrow extending beyond its stationary limits, these worms began extending themselves into realms not available to touch or taste. In the wisdom of the unconscious language, they met the future head on.

Organisms which could see commanded tremendous advantages over those which could not because, through sight, they

---

*Such signalling is still ubiquitous in the plant and animal kingdom and survives in the pleasant messages of the sex pheromones of the human female.

could learn of impending changes while the causes of those changes were still at great distances; they could react to conditions which were otherwise unpredictable from past experience of periodic change. Seeing organisms came to command foresight, and a type of future unavailable to creatures communicating by touch, smell, or hearing opened up to sighted animals.

Light sensitive regions are useless, however, unless they are part of an evaluation system. Hence, from its very origins, sight (as well as hearing) must have included not only organs of excitation and of action, but also processing facilities, recognized today in the functions of the central nervous system and the brain. Through its immense journey, the necessary now of life gave rise to the microstructure of time perception and matured, eventually, into the powerful features of the mental present of man.

6   Growth, Decay, and Organic Evolution

Life is ordinarily thought of as a one-way thrust, a syndrome of complicated processes, which lift matter out of its inorganic form and, for limited periods of time, make it display certain peculiar capacities, such as reproduction, aging and death. But we found that reproduction is not a necessary corollary of life, neither do all organisms age. Common to all life forms, however, is the unresolvable conflict between growth and decay. For the primordial, crystal-like forms of life, and even for cells of bacteria, this conflict is sotto voce; it is perhaps no more than a sophisticated embodiment of the opposing trends which we identified in the physical world. But once we begin considering eucaryotic cells, the growth and decay of life are amply manifest.

In this understanding of life, growth and decay cannot be given separate meanings; only their coexistence has ontic status. When we say that a child grows we do mean, of course, that he is getting taller, stronger, or her figure gets filled in. But in a deeper sense what we mean is that the life processes (growth and decay) are going forward: the child is very much alive. When we say that the dead body of a person decays we do mean, of course, that it returns to dust. But in a deeper sense what we are saying is that the life processes have ceased to function. Life lasts only as long as the conflict lasts, and hence, on the integrative level of life, the conflict is unresolvable.

While the body is alive it functions as does matter in general but, more importantly, as living matter in particular. If life is regarded as a manifestation of certain unresolvable conflicts, then living matter may be described as the intimate environment of life.

We found earlier that there are principles in the physical world which tend to keep the rate of disorganization (or entropy increase) minimal. With biogenesis, nature invented a method that can do more than minimizing decay rates: it can actively reduce entropy.* We have identified here an important continuity between non-living and living matter. To wit, the opposition to entropy increase continues across the eotemporal-biotemporal interface, but becomes much more efficient.

This crossing over is not without its discontinuities; we shall deal with them later. Here I wish only to draw attention to an uncertainty principle which is isomorphic with the one we found in physics. Both occur when we seek to connect features immanent in one level of integration with their origins on a lower level. Qualities well defined on a higher level may be ill defined on a lower one. For instance, what can we mean by the birth of the whiskers apart from the birth of the cat? Whiskers can grow, but they are not born. One can hammer out an answer but, figuratively, it will have fuzzy edges as do the momenta and positions of elementary particles.

Returning to continuities, we note that the action-reaction principle of the deterministic eotemporal Umwelt opens up to a "mutuality of adaptation" on the integrative level of life.

---

*Growth, evolution, and development are ideas one would prefer to identify with gain and not with loss, but we are hampered by the awkward terminology of thermodynamics. Because of historical reasons entropy is used as a measure of disorganization. Organic growth, then, amounts to loss, whereas decay amounts to gain ... of entropy. The whole thing began, said little Sam, when Billy hit me back.

Mutuality between organism and its environment is implicit in the Darwinian interpretation of evolution by natural selection. To each modification of an organism, produced by the selection pressure of the environment, there must be some corresponding modification in the environment. Predator-prey, or plant-herbivore systems, for instance, have been studied; they are often taken as single systems that show stability, oscillations, growth, or decay. But there is, of course, similar interdependence between life in general and its non-living environment, a subject of concern to ecology. Sea birds which filter out fresh water from sea water and excrete concentrated brine modulate the salt distribution of the ocean, even if ever so slightly. All adaptive changes in living organisms constitute selection pressures upon the environment, both organic and inorganic. We may indeed remark with Omar Khayyam:

> And strange to tell among that Earthen Lot ...
> "Who is the Potter, pray, and who is the Pot?"

We found earlier that with the steadily increasing cyclic spectrum of life, organisms came to generate new cyclicities to which nothing prior in the external world needed to correspond. Through the mutuality of adaptation these autogenic (self-generated) cycles came to exert selection pressures upon the environment. Thus, for instance, the feeding and reproductive cycles of organisms came to place certain demands on their environment to which the environment had to adapt. In its most sophisticated form, we may even think of the cyclic demands made by social institutions upon animals and plants.

In the mutuality of adaptation the thermodynamic arrows of growth and decay are as intricately intertwined as they are in the case of individual organisms. It is mostly the existential stress of life which we recognize. Its form is that of the irreducible difference between the expected and the encountered. The migrating geese over James Bay wending their way toward better climates, or the sad monkey pressing its lever for food, express the selfsame spite "to take up arms against a sea of troubles, and by opposing end them."

Focusing on the mutuality of adaptation and upon the existential stress of living matter is useful in elucidating certain issues of organic evolution and time. For instance, let evolution be defined as a creative process of successive forms of existents and levels of integration, such that each form or level may be associated with the idea of "more advanced" than the form or level preceding it. This definition presumes the observer's sense of time in terms of which successiveness can be judged. It also assumes that it is possible to distinguish among successive forms of integrative levels. Consider now that the distinguishable changes of physical cosmology, organic evolution, and the development of human knowledge are quite far apart in cosmology, closer in organic evolution, and may be quite close in intellectual growth. Evolutionary developments tend to follow one another more rapidly as one goes from astronomical objects to geology, to tree, to monkey, to man.

This apparent urgency in nature lures us toward a teleological interpretation of evolution in general, especially of organic evolution. The fulcrum of the arguments on goal-directedness is the question of time available vs. time necessary. Surely, no convincing case can be made for the coming about of even a flea by random shaking of chemical elements, even if the length of the shake were that of the age of the known universe. It seems as though the time that was available to create fleas is many orders of magnitude less than the time that would have been necessary for undirected evolutionary development.

Were the dynamics of evolution such as to favor, from among several adaptive measures, those which work the fastest, this would lead to creative evolutionary changes more rapidly than could random events. The effects of such a policy would be indistinguishable from the workings of teleological principles. But this is exactly what is happening.

For, consider that organic evolution does not happen in some evenly flowing, absolute Newtonian time. Were this the case, the sensitivity to such a time might conceivably be expressed in rates of evolutionary change. The situation, however, is the reverse. Organic evolution creates the operational properties of what we recognize as biotemporality. Living matter, throughout the history of life, created nowness, final causation, polarization of futurity and pastness, complexification, decay, etc.; that is, it generated all those features that determine the time of life. Therefore, it is not surprising that the career of organic evolution displays what appears to be goal-directedness. After all, it is that goal-directedness which (among other hallmarks) we identify with the time of life.

It is generally supposed that Darwinian selection necessarily implies a quantity called adaptation, or fitness, and that this quantity is bound to increase through evolution. P. T. Saunders and M. W. Ho argued recently that it is neither fitness nor adaptation that give direction to evolution but the increase in complexity, and this increase is a consequence of the way in which self-organizing systems optimize their functions with respect to local conditions. Their arguments are isomorphic, though not identical, with the ones given earlier. The conclusions are similar except that our primary interest is time. We may now spin this interesting story even further, in terms of the theory of time as conflict.

We have found that each adaptive step in the cyclic and aging orders of life inevitably produces increased existential tension. The more complex organism has more ways to fail than the simpler organism, even though the more advanced one may represent a better fitness. The most perfect adaptation is death: after a brief period of time, following death, the dead bodies of saints and sinners become indistinguishable from the dust of the earth. The adaptation is complete. Why is it, then, that organisms prefer to adapt by survival rather than by death? That individual members of a species sometimes do promote the survival of the species by dying rather than by living (in sociobiological jargon, by being altruistic), cannot be an answer. For, one can immediately ask: why does not the species adapt by dying out? And if one quotes examples of species that died out and thereby promoted the survival of the family, the kingdom, or even life at large, one can again ask: why does not all life on earth achieve rapid and effective adaptation to the inanimate environment, by perishing?

That living matter prefers to adapt by surviving rather than by dying is the great central dogma of evolution. As all dogmas it is taken so much for granted that even questioning it might appear senseless. But it must be questioned, because the answer to it is not at all self-evident. Yet it may be had from the conservative scheme of evolution by natural selection, provided it is admitted that the result of environmental selection pressure is complexification, whereas death is a return to less complex modes of existence.

As life evolved, the difference between the expected and the encountered was apt to increase, both for individuals and for species.* The difference between, on the one hand, what is possible to achieve by means of organic evolution, and, on the other hand, conditions for which natural selection was to select a fit, was apt to widen. The environment itself to which organisms had to adapt became more complex by the very actions of organic evolution. Consequently, existential tension had to become more intense. It follows that living matter was, and is, on a wild goose chase, because those measures of change that are supposed to narrow the gap between the expected and the encountered, lead only to a widening of that gap. Carrots, rabbits, and monkeys have no way to go, but toward a new integrative level, that can help life catch up with the conditions which its own evolution has created.

---

*Quantification of various needs of an organism amount to straightforward measurements of existential tension which, in more general terms, is the difference between the expected and the encountered. Thus existential tension is more readily quantifiable than are fitness or adaptation.

V Brain, Mind, and the Self

PERHAPS as long as fifteen or as recently as three million years ago, in the course of organic evolution, in one or more members of the Hominid family the organization of the central nervous system crossed a certain threshold of complexity. A portion of the new nervous system was the ancestor of the human brain. At some period during the eons between the appearance of the human brain and that of the first human groups, the brain of genus Homo assumed a number of control functions which, in due course, set man's behaviour distinctly apart from all animal behavior. This fateful change, which commenced only perhaps 100,000 years ago, made it possible for man to work out future behavioral strategies for imaginary conditions, based on past experience. In this process an enduring entity, the self, came to be defined, and a unique method of communication developed, based on the symbolic transformation of experience.

The reasons for the peculiar behavior of man have been attributed to different causative agents in different historical epochs. Contemporary use would describe the peculiarities as mental phenomena and would attribute them to the functions of man's mind. What we have been calling the nootemporal Umwelt corresponds to the integrative level of the mind. In this chapter we shall explore some of the features of the nootemporal world.

1 Expectation and Memory

No special arguments need be given to stress the tremendous adaptive advantages that accrue to an organism if it can prepare for future contingencies on the basis of past experience. Although the capacities of expectation and memory necessary to accomplish such feats are immensely more developed in man than in other organisms, they are certainly not limited to our species. Consider, for instance, the dance of the bee.

Let us assume that the dance, whose purpose is to entice others in the hive to foray to an area which the bee has found and upon whose location it now reports, may be translated to mean something like "I shall fly to a place which is at ..." Other bees observe and imitate the dance and gradually change the message to a collective statement: "We shall fly to ..." We cannot make sense of the bees' behavior (going to the place identified by the scout) without explicit reference to time. We must assume that the creature present in the "mind" of the bee includes some dim, practical knowledge of continuity, a primitive belief in a future that can be acted upon from experience gathered in the past.

Let us regard this hazy knowledge of the past as an instance of recall and distinguish it from memory. Let *recall* mean a response to external stimuli in a manner that indicates that the stimuli have been encountered before. Thus, Odysseus was recognized through recall by his dog Argos, after twenty years of absence, even though he must have looked different from the young Odysseus. Argos acted upon this recognition by barking. Likeweise, the bees of the hive recall, as it were, the dance of the scout, although in detail it is likely to differ from prior dances, and act upon that recall by going on their foray. By *memory* I mean a re-presentation of past experience in response to stimuli which themselves are only contingencies and not actualities. Imaginative planning of a journey in mountains which I have not seen for many decades must derive from memory. One may have memories of memories and of expectations, and also memories of past recalls, but it is not possible to recall earlier recalls. By the principle of parsimony, until evidence appears to the contrary, we must, therefore, assume that bees have a capacity for recall: the dance is part and parcel of their trips out to and back from the hive. The ancestry of recall includes physical forms: even magnetic cores are capable of recall (even if it is named "memory"); the ancestry of memory, however, is limited to man and the most advanced animals. The behavior of sleeping dogs suggests, for instance, that they dream; if true, this demands rudiments of memory.

The origins of man's phenomenal capacity of memory are

unknown but, let it be assumed that for some reason the expectations of our distant ancestors began to extend increasingly into the future so as to include, eventually, the certainty of personal death. To the capacity of displaying mortal terror in face of present danger, a behavior that man shares with advanced organisms, there came to be added a new knowledge that of death. This new concern is likely to have urged our ancestors to explore the past for cues of future expectations. The length of time into the future through which consequences of present actions could be traced, came to be related to the ability to organize mentally the memory of past experiences.

As with all adaptive steps, improved fitness brought with it increased existential tension. Whereas the behavior of animals was directed toward securing satisfaction of immediate or cyclic needs, in the case of man there was added the capacity to manipulate images of the past. This manipulation was spurred on by a knowledge of inevitable death, disturbing and hence, in some ways, better to be forgotten.

There are goods reasons why in this speculative reconstruction I have placed our concern with futurity before that with pastness, and both before the mental present. This sequence of development occurs ontogenetically, phylogenetically, and even in the growth of civilizations.

Consider that very soon after birth, perhaps even in utero, the infant's physiological present begins to open up, under the pressure of anxiety and expectation. Post-partum food is not supplied free of charge; it must be reached for, and satisfaction is often delayed. Those who have examined this issue hold that the earliest extension of future horizons in the growing infant is tied to the anticipation of food in general, and of the breast in particular. Already at this stage, anxiety and a primitive knowledge of time reinforce each other.

Words relating to time become meaningful for children in a fairly universal sequence, regardless of the cultural milieu. At about 18 months American children can respond to "soon" as a time concept connoting the future, by waiting. They can use "soon" in a phrase by 24 months; they begin verbalizing spontaneously the word "tomorrow" at about thirty months, but only later are they able to use "today". Mastery of the word "yesterday" tends to come last.

This sequence is the same as are the steps in cultural development. Preparations for the immediate future, such as for the coming evening, are surely very ancient habits of man and beast; but preparing for a longer future, based on the memory of past experiences, is demonstrated only (through the development of tools) in the upper Paleolithic period. The emerging apprehension of long-term future and meaningful past are tied to the evolution of symbolic thought, and possibly stem from an ability to imagine events and confer upon them a sense of future reality. Selection among imagined events must then be made, and the possible, the probable, and the impossible distinguished with the assistance of the mental imagery of past experience. It is man's ability to manipulate his imagination of future and past that makes his behavior plastic, hence fundamentally distinct from the behavioral rigidity of even the advanced primates.

Very short term memory (up to about 0.5 sec), short term memory (up to about 20 sec), and long term memory can be shown to involve three distinct mechanisms, but we have no firm knowledge of how memory is retained. Nor do we have an answer to a fundamental question related to time and the brain, formulated by the British neurologist, Lord Brain. If memory corresponds to a brain state, expectation to another, and sense impression to yet another, and these three are simultaneously present, how may they be distinguished? I shall try to develop a speculative answer to this question.

Brain states that represent past conditions must contain some recognizable contrasts and it is these contrasts that must be stored as memory images.* But if so, then the contrasts themselves

*"Mene, Mene, Tekel, Upharsin," carried its prophetic message, hence constituted a memory storage, because the written words constrasted with the unannotated wall. Likewise, in a visual impression, where there was no image on the retina before, now there is one, or vice versa.

(mechanical, electrical, or chemical) are the obvious candidates for carrying the temporal coding, for they are, by their very nature, polarized. The suggestion emerges that the simplest temporal labeling of a memory image may be diadic, that is, a coding of before-after without a now. This would make memory images eotemporal. Since pure succession is asymmetrical, as it were, in both directions, one would think that an eotemporal world will appear to the conscious mind as stable and yet plastic. Introspective evidence suggests that the Umwelt of memory is, indeed, eotemporal.

Consider the curious static and modular character of memory in man. Memory images do not form a continuity comparable to the successive frames of a motion picture film; they are more like snapshots in a family album, with images overlapping helter-skelter and in need of information from the photographer. For instance, I can remember Sandy running down the hill toward three copper beech trees. Though her trip took perhaps a minute, it appears in my memory in all its details simultaneously, as a unitary static module, containing motion. I also remember that she reached the trees and slowed her motion in the warm shadow of the summer sun; again a static, unitary module that includes motion. If, instead of beholding these stationary tableaux of change, I wish to remember the grace of her motion through the grass, then remember how she embraced beneath the tree a man younger than I am, then I must make an effort to integrate the separate images. I must force myself to slow them down; I must force these diadic, eotemporal projections into my nootemporal world. I must attach these images of pure succession to the futurity and pastness of the mental present of my conscious experience.

By comparison, let us now consider images of the future. I can imagine myself as the lover of Liv Ullman, but I feel no need to control my imagination through temporal restraints. The individual "snapshots" of this affair appear to be as static as those of memory, but they bear not even an eotemporal labeling. They have a freedom, an undeterminateness, that may be characterized as prototemporal.

Where do these introspective facts leave the issue of memory, expectation, mental present, and the brain? It seems to me that memory images exude what I shall later describe as eotemporal "moods", but with a bit of directedness, as though they were about to enter the biotemporal world. Images of expectations determine an even lower Umwelt, for they hover between totally disjointed and somewhat organized "snapshots." As sense impressions, working in the mental present, remove some images from the store of expectations and add them to the store of memory, they also attach to them a temporal labeling. This, I believe, is a selection process taking place in that metastable region between mind and body that we know as the unconscious. We shall have more to say about this later.

It is reasonable to assume that memory evolved from such instances of repeated recall as were reliably consistent features of the world. For instance, a decrease in light intensity might be predictably followed by a drop in temperature. Surely, it was not the cue that had to be of importance to the organism but the change that followed the cue, in this case, the drop in temperature. It would seem, then, that beginning with modest anticipations tied to present cues, organisms evolved to anticipate an increasing variation of cyclic changes and eventually, of noncyclic changes also, as long as these were reliably repeated by the environment.

The stance just expressed is one of greatly expanded scientific empiricism, combined with the extended Umwelt principle. In this view the source of all knowledge (the store of expectations, memories, and sense impressions) is experience. But the basis of this empiricism is not the individual experience alone, not even the experience of the human race, but that of the evolutionary chain of life and matter. If we were to attempt to trace the ultimate sources of all human knowledge and tried to identify our earliest collective memories, we would find ourselves regressing from the nootemporal to the biotemporal, thence to the eotemporal Umwelts of organic evolution and facing, eventually, the issue of Creation.

Be that as it may, organisms evolved so that they could foretell the behavior of friend, food, and foe. It is not a mere figure of speech that advanced organisms, with faculties of expectation, recall, and memory, carry within their bodies internal charts of their external landscapes. The behavior of the actors upon that landscape, at least to first approximation, is predictable and independent of the organism itself.

2   The Self

In its earliest forms, anticipation and recall had to be based on an image of the world in which the behavior of the external world, by and large, was quite independent of the organism itself. Thus, darkness was predictably followed by a drop in temperature, and the spectrum of the sun remained the same, day after day. As living things improved their capacities for anticipation and recall, such environmental predictability ceased to be the case, at least as far as the living environment was concerned. Prey learned to evade the predator and foe learned to circumvent the defenses of foe. The inner chart of the external landscape, recorded in the biological and the nervous organization of animals, had to allow for those actions of other organisms which were in response to the movements of the animal itself. In due course the set of symbolic actors on the inner stage had to be enlarged so as to include a new symbol, one which could be held responsible for the responses of external objects. In its most sophisticated form, in the mental world of Homo sapiens, this new object is the self of man.

The assertion that such an evolutionary development did take place derives entirely from the logic of the situation. Dating this event is a very difficult task. The first stage in the evolution of modern man commenced some 10 or 15 million years ago. Next came a long period of slow change, perhaps 5 million years, followed by perhaps two million years of speciation; then came an

Eva de Maizière, *The Mask*, 1976. Courtesy of the artist. Bonn–Bad Godesberg.

OUR IMAGE of the self represents an object which is assumed to function in time and space as do all other objects. But unlike other objects, the self can only be partly apprehended through the senses; partly it is a symbolic continuity to which nothing in the external world corresponds.

> Doe a deer, a female deer,
> Ray a drop of golden sun,
> Me a name I call myself
> Far a long, long way to run ...
>
>   Rodgers and Hammerstein
>   *The Sound of Music*

accelerated increase in brain capacity, the change to erect posture
and bipedal locomotion, and the evolution of the hands. Perhaps
100,000 years ago, building upon the genetic potential that
evolved during the prior millions of years, the brain seems to have
crossed a threshold of complexification, making the psycho-social
evolution of man possible. There was a continuous growth in the
relative size of those regions of the brain that were not given over
to registering incoming stimuli or coordinating outgoing responses, but were retained for the pursposes of a "map-room."
Somewhere toward the end of this vast trajectory evolved man's
capacity to become an individual.

The self is a unique and curious symbol. It represents an
object "I" which is assumed to function totally in the external
world as do all other objects, yet it cannot ever be totally seen as
can other "selves." It cannot be properly explored by smell, and
only partially by touch. As an object it is only partly sensate;
partly it is an imaginary construction as was a small bear hiding
underneath our steps: it could be heard every evening, but it
could never be seen.

Though I cannot completely explore myself by sight, touch,
or smell, I can hear myself almost the same way as other people
hear me. The intimate connection between hearing and the self
makes one suspect that selfhood emerged only when, in the course
of evolution, the auditory modality became a significant source of
information. It is believed by some who studied brain casts of old
skulls that in early mammals hearing first evolved as an analog
and assistance to vision. If this was the case, physiological correlates would have to have included the need for enlarged neural
complement in the brain, and an increased use of temporal coding
to assist the spatial coding of sight. The intimate connection
between language and selfhood, between naming and identity is
well known; the absence of language might thus be the most
important single reason why even hominids did not evolve
identities comparable in importance to that of the identity of man.

An organic or biological know-how of separating permanence from change, such as is evident in animal behavior,

surely develops in the human embryo while it is still in the womb. Perceptual separation of the self-as-permanence from the non-selves begins when the personal cosmology of the infant commences, that is, when he leaves the darkness of the womb and first sees the light. He learns to distinguish, mainly visually, certain patterns of permanence in a world that is otherwise in ceaseless flux. Within the class of permanent things he learns to identify objects in motion, based on a smooth pursuit system of vision, and objects that are static, based on the saccadic system of his eyes. It is at this early developmental level, and in the structures which correspond to them in the psychobiological organization of the adult, that we must seek the origins of separation between being and becoming, and even the sources of Zeno's famous paradox of the flying arrow.

The infant begins to build theories about the world. The later, mature self is therefore not simply a store of memories of objects and events, but also one of memories of theories about reality: smells and voices that are good and bad, mother-things that soothe or frighten, and even a belief that the body which fell asleep last night is the same as the one that awoke this morning. I shall argue later that it is this theorized, or discovered continuity of the self that constitutes the paradigm of one-ness, hence the necessary fundament of quantification and number.

The idea of the self makes it possible to work out complex strategies in imagination: if I do this, foe shall do that, friend will do such, food will run so. A sophisticated scheme for generating predictions about the world thus becomes possible, including predictions about the behavior of the object that is "me."

Let us now imagine the drama of selfhood played by two actors; they are, of course, only projections of a single, functioning human identity, but we shall find them rather useful in our analysis of the mind. We shall further imagine them as functioning in different temporal Umwelts. The main task of one actor is self-prediction: his Umwelt is nootemporal, he contemplates a future different from the past, and may boast an ill-defined (mental) present. This actor is the *Agent*. The Umwelt of the other actor is that of memory images, expectations, and dreams. It is a conservative, eotemporal world for it is without a now, and therefore, without a clear distinction between past and future. We shall call this actor the *Observer*. The behavior of the Agent may be associated with that of the conscious self of historical man. The behavior of the Observer corresponds to the paleologic of archaic man, surviving in the mind of modern man.

Since the Agent and the Observer are only analytical portions of a single selfhood, they must be thought of as existing in symbiosis: they both get the same sense impressions and, we must assume, they communicate. When it comes to issues of survival, the Observer's paleologic will make its own suggestions for preferred behavior and, in retrospect, some of its suggestions might turn out to have been correct. But such guidance as appropriate to the eotemporal Umwelt may easily appear as ill advised to the Agent; the Agent should draw upon, but not trust implicitly, the opinions of the Observer. After all, though mankind is moved by its dreams, the world is not one of eotemporality – at least not as seen by the Agent. On his part, the Observer must judge the Agent's mode of thought and action as fumbling, and often, as unintelligible, for the Agent employs nootemporal categories of which the Observer can know nothing.

In the integrated structure of the self, of the "me," we witness and experience the struggle of two different interpretations of sense impressions: they are two different mappings of the external landscape into the internal one. One corresponds to the older evolutionary mode that resembles the eotemporal, the other the newer mode, the nootemporal. It has been shown in studies of attribution that when we act as observers (of another person) we tend to assign the causes of his actions to stable personality patterns; when we are the observed and we speak about ourselves, we tend to attribute our actions to the reality of situational requirements. Indeed, the Observer-self seems to prefer things stable (more of the mood of the eotemporal) whereas the Agent-self prefers things dynamic (more of the mood of the nootemporal). These preferences may be identified in the traditional personalities of the Prophet (the Observer) and the Statesman (the Agent).

If the world of the Observer (Prophet) is one without temporal direction, whereas that of the Agent (Statesman) is one of futurity, pastness, and presentness, can the Observer foretell events in the Agent's future which the Agent cannot know? Can he, in the traditional sense, prophesy?

The answer is an unambiguous "sometimes yes, sometimes no."

To the noetic Agent the words of the Prophet sound as though possessed of a spirit or divinity. He has charisma (from the Greek, meaning "gifts," usually divine) and may even be speaking in outlandish tongues. The Prophets's voice is that of the paleologic of the mind. But if so, then the temporal Umwelt of the Prophet is not that of a superhuman godhead who knows everything (hidden from the eyes of the Agent) but that of a child or advanced animal that lives in an eternal present. Just as the child knows with absolute conviction that this is this and such is so, so does the Prophet know that fate is fixed, events preordained, the future predictable. In a way appropriate to eotemporality, the causation of the Prophet's world is deterministic. "Alone he saw the field of time, past and to come," says Homer of Mastorides, the Forecaster.

This primititivenesss endows the Prophet with certain advantages. Perhaps time was invented so that we may not realize what we, in fact, do know about the future. Listening to our inner voices over the din of everyday life can conceivably reveal predictable details of the future which our busy and fearful Agent (Statesman) could not admit. Such prediction must, however, remain random and, until the event, unprovable. Were it otherwise, we would get into the vicious circle of time travel that would make any prophetic enterprise intellectually dishonest.* In this

---

*Prophesy a future event. (a) Take steps to prevent its coming about, and it does not come about. It follows that the prophecy was false. (b) Find that it is impossible to interfere with fate and in spite of, or because of, your actions, the event does come about. This is a selfconsistent condition but its total determinism is difficult to accept.

Time-travel into the past. (a) Misdirect your grandfather so that he will never meet your grandmother. You do not exist. (b) Find that it is impossible to interfere with the past. You are then a historian, not a time traveler.

view prophecy is one of the creative capacities of the mind. Its occurence is just as intrinsically unpredictable a priori as was the genius of a child, born in the house of Vice-Kappellmeister Leopold Mozart on January 2, 1756.

In philosophy as well as in literature, personal identity has been recognized as a troublesome and elusive idea even somewhat mysterious. The elusiveness is not surprising if it is recalled that the self, like all identities, is said to function in the external world although, unlike any other identity, it is only partly on the external landscape. That there is, indeed, something quite peculiar about this object, and that this peculiarity pertains to temporality, may be demonstrated by trying to find an answer to the question of what we mean by the self in the context of death?

It is not difficult to interpret what is meant by "I" in the sentence "I don't want to die!" if the shout is coming from someone before a firing squad. The "I" is the prey, the "not I" the predator. But what if the cry comes frome someone dying of metastasized carcinoma? There the spirit is unwilling to perish while the body relentlessly decays to death. In suicide the roles are reversed: "the spirit indeed is willing, but the flesh is weak."* In all three examples the symbolic construct, the self, insists on remaining the executive power: it is *its* will that ought to be done, not that of the non-self. In ritual and in tradition this power and importance of identity and its relation to language has been widely acknowledged. In certain civilizations name giving is delayed until the viability of the infant appears certain. Yahweh, the proper name of the God of Israel, probably standing for His claim "I am who I am" was judged too sacred to be uttered, except on special occasions.

Information on the self has the character of privileged disclosures: details must be reconstructed from overt behavior. It remains privileged, and a bit mysterious, even for the body of

---

*In this tragic utterance "weakness" is the strong opposition of the body to the designs of the spirit.

which it is said to be the self. Consider, for instance, the hero ready to sacrifice his life at the peak of good health. Why should the body not revolt against such murderous designs, drawn up by an alien king, even if he rules, presumably, from the body's own head? Or, in case of a painful disease, why should the self insist on survival if the body is ready to perish? This foreign ruler giving orders to one's body is hidden behind the palace walls; it is difficult to find out about one's "real self." "In many ways the saying 'know thyself' is not well said", wrote the Greek dramatist Menander some twenty-three centuries ago. "It were more practical to say 'know other people'."

Through the use of the internal landscape, refined by the inclusion of a symbol for the self, our capacities for predicting the behavior of others have become highly developed. Our daily life is conducted in the belief that, in many ways, the future shall copy the past. In cultural anthropology one even speaks of the "generalized other," that is, of a set of standards that the individual attributes to others so as to predict and interpret their behavior. But expectations concerning the future behavior of our own selves have retained a degree of uncertainty; there is no such thing as the "generalized I."

Donald McKay has given several isomorphic arguments in favor of the idea that a measure of uncertainty is intrinsic in the self-prediction of all systems which are sufficiently complex so as to be able to make non-trivial statements about their future behavior. From our point of view, it seems that although the self has been successful in learning to predict the behavior of foe, friend, and food, it has retained a degree of indeterminism about self-prediction.

E. O. Wilson in his far-ranging study of *Sociobiology* remarks that at the upper end of the evolutionary ladder, typical vertebrate traits include refined communication and personal recognition; as a consequence, a typical vertebrate society favors individuals at the expense of social integrity. Why, he asks, in spite of this essentially "destructive trend," human societies do not fall apart but can achieve an extraordinary degree of cooperation, is "the culminating mystery of all biology."

Wilson has properly sensed that selfhood is a corollary of being human, a fact that has been current in philosophic thought for a few millenia. The answer to his question, if it is sought within biology, must remain a mystery – in that sense of mysteriousness that we shall discuss later, in connection with level-specific languages. The issues may, however, be elucidated if we leave the integrative level of life and consider the cohesion of the nootemporal world, that is, the integrative level of the mind.

3   The Mind

The tremendous advantage conferred upon man by selfhood is connected with the functions of his central nervous system and, especially, with those of his brain. But what specific quality of his brain ought to be the criterion for the step to manhood?

Quantitatively, the most obvious parameter which suggests ifself is that of size. Studies of the endocasts of fossilized mammals (casts of the inside of the cranial cavity) show that as long as 65 to 30 million years ago, their brains began to increase in volume at higher rates than would be expected from their lineage and growth of body size. Endocasts examined by H.J. Jerison suggest that about 50 million years ago fissurization evolved, possibly as a way of increasing the available cortical surface (as compared with smooth brains of identical volumes). In the evolution of genus Homo, the first substantial enlargement began perhaps five million years ago with the cranial capacity changing from about 300 cm^3 to 4–500 cm^3, reaching the sapient level of up to 2,000 cm^3 about 200,000 years ago. There is no similar change in the encephalization in any other mammal during the last five million years.

The difference in mental capacities between man, and whatever might correspond to mental in the behavior of the most advanced primates is so vast that it is unimaginable how an increase in brain volume alone could possible account for it. It is much more likely that what matters is the kind of brain that an organism has, and not how large it is; indeed, the information processing activity of a unit cortical volume in different species is independent of the brain size of that species. "It does not matter in the least having been born in a duckyard," said Hans Christian Anderson's Ugly Duckling, "if only you come out of a swan's egg!" Thus, nanocephalic dwarfs (dwarfs with skeletal proportions of normal adults but seldom taller than three feet) have brains whose volume is a third or less than that of a mature adult; still they acquire verbal skills which apes of equal or larger brains cannot do.

For the biological basis of man's phenomenal creativity we shall have to look at some features of the brain other than its mere size. We may take our cue from the great British evolutionists, C.S. Sherington, who held that man's very complex behavior must be accounted for in the complexity of the synaptic mechanism of his brain. We shall approach this by first clarifying some ideas about the mind.

When we considered biogenesis, I spoke of life as functioning in the environment of inanimate matter. "Environment" meant more than it does in the sentence "the environment of the polar bear is snowy," because it included the body whose life was being considered. The environment of the horse's life is the pasture as well as the matter that makes up the horse. That portion of matter which is most intimately related to the life of the horse, its soma ("another experiment of the gene") is organized in ways peculiar to living matter.

Likewise, mind operates in the environment of living matter. But the living matter that makes up the immediate environment of the mind functions in ways that are peculiar not simply to living matter in general, but to living matter that thinks. This peculiarity is the *complexity* of the human brain in general, and the necocortex in particular.

Complexity, as here understood, is not a measure of the number of component units of a system, but of the number of distinct ways in which the members of that system are, or may be, interconnected. Consider, for instance, that the number of neurons in the human brain is of the order of 10^{10}, and that each neuron has some 5,000 synaptic contacts, making the total points of possible connections 5×10^{14}. This is a very large number, but quite negligible in comparison, for instance, with 10^{22}, the number of molecules in a cubic centimeter of water. But these 10^{22} molecules cannot be arranged to form many distinct states; their physical and chemical properties do not differ in any important way from those of 10^{37} or 10^7 water molecules. However, the number of possible, even if not actual, connections among synoptic contacts in the brain is of the order of 5×10^{23}. And even this is only the number of "wirable connections" and not the number of possible states of such a network.

Stafford Beer calculated that, as a digital device with each axon either being on or off, the number of various brain states is about 10^{10^9}. This number may be described metaphorically as countably infinite even if Beer made a mistake of a few billion. By way of comparison, consider that the number of different sequences of possible moves from the beginning of a game of chess, played on 64 squares, in the calculation of C. E. Shannon, is only around 10^{120}. The number of electrons in the universe is estimated as a mere 10^{79}, and the number of different sequences of a checker game as a diminutive 10^{50}.

It is rather an understatement that the variety and number of possibilities in the physical universe staggers the imagination; I would not know how to compute or even what to mean by the number of its possible states. But if we identify the number of potential brain states with that many engrams (10^{10^9}), then the wealth of imagery that the human brain can store should stagger the universe. Nowhere in the physical world do we know of a complexity even distantly comparable to that of the human brain, except other brains.

The "countably infinite" number of brain states does not in itself vouch for a meaningful connection between them and the creative capacity of man; for that, brain states and creativity must be connected by a convincing theorem. A rather crude and not altogether convincing one might be derived from a theorem by W. C. McCullagh and W. Pitts which states that "anything that can be completely and unambiguously put into words is ipso facto realizable by a suitable neural [relay] network." Perhaps this argument can be turned around claiming that a relay network of 10^{10} neurons permits a large variety of unambiguous as well as ambiguous, verbal as well as non-verbal statements.

A less rigorous, but more impressive reasoning is due to E. O. Wilson. He notes that a minimal set of essential responses to token, guiding stimuli from the environment, found in the most primitive instinct-reflex mechanisms of sponges, coelenterates, and flatworms represents the action of a central nerve cord of hundreds or thousands of neurons. An organism with a fully elaborated central nervous system and a brain containing 10^5-10^8 neurons can learn to handle particularities in its environment (such as recognizing places); examples are lobsters, honeybee workers, the cold-blooded vertebrates and the birds. Organisms with 10^9-10^{10} neurons have the capacity to generalize and juxtapose patterns; in them, complex behavior is not totally programmed and socialization is prolonged. In man, the capacities include, in Wilson's words, the "perception of history." In our vocabulary, this is the knowledge of time.

Although I would not know how to compare the intelligence, or mind, of a sponge (10^3 neurons) with that of Lucretius (10^{10} neuron ï), the progression seems to run very far ahead of increase in brain size, or even of increase in neuron complement. It is more as though we are dealing with an increase of learning capacity proportional to the factorials of the available neurons or – to the complexity of the neural network.

Let us recall John von Neumann's reasoning that there is a degree of complexity below which, if an automaton makes another one, the new one will be less complex; above which, it is possible for automatons to construct more complex automata. We have already used this argument in connection with the coming about of life. Likewise, this is also what might have happened to the brain. It reached a threshold of complexity: above that boundary, using the symbolic transformation of experience, a mind can produce (that is, educate) other minds that may be equally or more complex; below that, it cannot. The region where the threshold is crossed is the interface between the biotemporal and nootemporal integrative levels.

Consider now that the very cold (absolute zero), the very hot (the primordial fireball), the very large (the universe), the very small (particles), the very many (probabilistic aggregates) and the very fast (motion near the speed of light) have their peculiar laws, even if the boundaries of their effectiveness cannot be precisely drawn. I believe that the *very complex* also has its peculiar lawfulness, even if there are no unambiguous limits through which one must pass to reach it. Our psychobiological organization and our size do not place us anywhere close to the very cold, very hot, very large, very small, very many, or very fast, but having a mind places us at the frontiers of the very complex. We have no evolutionary experience with the physical extremes (very large, small, many, fast); hence such conditions appear to us as rather strange; but we have been living with the very complex for millenia, or perhaps millions of years, hence it must be the case that the body evolved so as to benefit from it. If so, we ought to be able to identify some of the ways in which we cope with the very complex nature of our brains.

Consider, then, that the amount of time that would be needed by the theoretically fastest possible computer to trace all the possible sequences of a chess game (10^{120}) is of the order of 10^{80} years. This is a long period, considering that the age of the universe is only about 10^{10} years. The system of steps through which we would have to search is known as the "natural tree structure." One begins by moving one chess figure by one allowed step, and then considers all other allowed combinations. Then repeat it for all other possible steps of the same figure, then for all other figures. Such structures become computationally unsolvable when the points of branching approach 50 (corresponding to 10^{15} alternate possibilities).

That chess players can still play intricate games and, so far, have beaten all computer opponents show that it is not necessary to examine all possible states up to a Final Judgement. For instance, a chess master might make his judgment on the basis of the geometry of chess figures on the board and extend his tactics to only a few future steps. Instead of climbing along an actual tree to locate the single apple at the end of a single branch, one may take a photograph of the whole tree with a red-sensitive film and find the apple in one fell swoop. In both the chess and the apple-tree examples, a view of the system as a whole, at an instant, instead of analytical exploration of all possibilities, seems to be sufficient to yield, if not a rigorous solution, at least a mastery of the situation.

Let us now distinguish between state and process descriptions of systems. A state description is a family of static lists; each list gives the status of the system at an instant; it is a series of spatial specifications. A process description gives rules, instructions, or reports on the functioning of the system; it is a temporal presentation. For reasons of parsimony, process descriptions are of great advantage for complex systems. Natural selection, for instance, found it advantageous to instruct new life via genetic programming, rather than by genetically carried state descriptions. If life on earth reproduced by state descriptions, then each gene would have to carry specifications for each new state of developing organism, snapshots from cradle to grave. The rigidity and unwieldiness of such a schema is obvious.

When it comes to describing the functions of the human brain, process description is not only more parsimonious than state description, but surely, the only feasible one. I am unable to produce even a single state description of my brain, but I can give a process description of its functions through language and action such as, for instance, by writing these pages. Since the only feasible accounts of the states of the very complex are through process descriptions, the mind is purely temporal by necessity. We may even turn this reasoning around and say that the nootemporal Umwelt is determined by signs and symbols appropriate for the description of the immense complexity of the brain.

The inner map of this nootemporal world is a set of hypothesized traces in the brain, named engrams. We do not know what constitutes an engram, but that they do interact and combine is demonstrated by the existence of coherent languages. It will be recalled that when life first emerged, the broadening spectrum of the cyclic order began to generate new biological cycles to which no prior external conditions needed to correspond. Likewise, as the store of engrams increased it surely generated new engrams to which nothing in the external world needed to correspond. I shall describe such totally self-generated engrams, and the images and thoughts that correspond to them as *autogenic*. The capacity to form autogenic imagery may be identified with creativity in man; paraphrasing Pascal "the mind has its powers that the brain knows nothing of."

Among the autogenic images (loosely, in the imagination of man) there will be some that can be made to correspond to sense experience through appropriate behavior. For some others such correspondence can never be established because their realization would demand the negation of physical or biological regularities. Selection rules that assist individuals and collectives in deciding which of the autogenic images are to be taken seriously, and how they are to be acted upon, have traditionally been classed under three major headings: the true, the good, and the beautiful. We shall deal with these categories later.

I said earlier that the self-generated biological cycles of life exert selection pressure upon the environment of the species; in exact analogy, the autogenic imagery of the mind exerts selection pressure upon its environment, which is the brain. Through the central nervous system imagination directs the behavior of the body and through bodily functions, it tends to modify the living and inanimate environment of the individual and of the species. Cuneiform signs appear on slabs of stones; dams are built across rivers; people are ordered to kill, maim or glorify things, ideas, and other people. "What is now proved," wrote Blake in *The Marriage of Heaven and Hell,* "was once only imagin'd."

Autogenic images, memories, and expectations fill man's mental present. Unlike simpler presents whose limits were, more or less, expressible in numerical readings of clocks, the mental present has no definite boundaries. At one instant a person may be concentrating on the task at hand, the next instant he may conjure up his whole life and embellish his memories with expectations; yet at another instant the mental present might comprise only memories of certain hours.

Manageable order and unity are introduced into the hubbub of the mental present by that artificial narrowing of our interest which we call attention. The role of attention may be likened to the functions of the pick-up needle of a phonograph. That needle follows only a single undulating line, yet from its motion, communicated to the ear, a listener may reconstruct the sounds of distinct instruments playing in a symphony orchestra. Likewise, focusing upon a narrow present introduces continuity, order, and unity into the rich harmonies of the mental present. The narrowed attention to the present is a corollary of the sharp separation of memory from expectation, past from future.

There seems to be a reciprocity between the unity of conscious experience and the degree of focusing on a limited present. In healthy mental equilibrium there is an unquestioned sense of unity of the self; the loss of this unity is often a clinical sign of disturbance, accompanied by an incapacity to concentrate upon the present. Whenever our attention to the present relaxes, as in certain cases of ecstasy, so does our keenness about past, future, and personal identity.

The handling of autogenic imagery, as manifested in the way a man behaves and what he does, is closely tied to the balance between the roles of the two imaginary actors who play the drama of selfhood. In a mature individual the Agent and the Observer function in a unity of protracted conflict: they distrust each other's judgements, but somehow their two Umwelts complement each other for the benefit of the survival of the individual. Under great strain, or sensory deprivation, or in the case of psychosis, these analytic ghosts might actually be torn asunder. Because their temporalities are eotemporal (that of the Observer) and nootemporal (that of the Agent), respectively, their evaluation of the usefulness for survival of autogenic images may be quite distinct. What for the future-past-present world of the Agent is an impossibility might appear to the Observer, in his Umwelt of pure succession, as quite "reasonable" and even likely; thus plans that might take superhuman powers are judged feasible and a nation takes on the task of fighting the world.

I would like to identify the protracted conflict between the two functions, symbolized by Agent and Observer, with *conscious experience*. Since the mind subsumes the Umwelts of both actors, man has no absolute inner reference through which he can identify final reality; Agent and Observer often depict the world differently. The autogenic images that float in and out of the mental present may or may not have any correspondence to the external world. Hence, one has a continuous feeling that one's potentialities (which include impossibilities) are immensely greater than the possibilities permitted by the restraints of the organic and inorganic worlds.

In terms of sensual cognition, this feeling confers upon man a unique privilege: that of suicide. What with the struggle between Agent and Observer, the eotemporal endlessness of the Observer's Umwelt might appear more desirable to the self than the threats and promises of a nootemporal world of uncertain future. In terms of intellectual cognition, the same feeling of vast possibilities finds expression in the idea of *free will*.

Let us recall now the hierarchy of causations. In the atemporal chaos cause and effect have no meaning; in the prototemporal world they are probabilistically related; in the eotemporal world cause and effect are deterministic. In the biotemporal world final causation appears, defined in terms of the self-interest of the organism. In the nootemporal world, the free will of man makes it possible to enrich this hierarchy of possible connections with that of intentionality in terms of symbolic continuities. Free will permits man to select such courses of action as appear helpful for the survival not only of his body, but primarily, of his ideas and ideals. Thus, he can subordinate his individual life not only to the biological interest of his species, but to the interest and survival of the symbolic continuities of his mind. One should want to extend this hierarchy to the societal integrative level, that is, to the family of man on earth. But such a unitary society of man is only in statu nascendi, (perhaps), and the issues are hazy.

From an evolutionary viewpoint, we have already identified the primitive Umwelt of the Observer with the paleologic of the mind, probably surviving in the function of the paleocortex. The functions of the newer regions, those of the uniquely human areas of the cortex, probably correspond to the Umwelt of the Agent. The conflict between the two modes of thinking is indigenous to the mature mind and must remain unresolvable. The self lasts only as long as the conflict lasts between the older and the newer Umwelts, between knowledge felt and knowledge understood, or briefly and poetically, between passion and knowledge. Separately and together, these modes of cognition generate the imagery of all that man can expect to find or achieve. The difference between the expected and the encountered, as in biology so in the world of the mind, is a form of existential tension. As a semi-autonomous organism whose many levels retain the hallmarks of their evolutionary development, the mature individual is motivated by a hierarchy of existential stresses.

Whatever conflicts we identified for matter certainly apply to man; the dynamics of a rolling human body is the same as that of a rolling stone. In the biological world, the quality of the stress experienced by a hungry man is not different from that of a hungry horse, frog, or eagle. The radical jump comes with the generative powers of the mind which finds that the fulfillment of its imagination is restricted by the biological and physcial laws of the world.

Earlier I reasoned that the increasingly sophisticated adaptation of evolving life has inevitably led to a continuously increasing existential tension. Identical fate befell the mind. While it succeeded in improving man's capacities of adaptation (until the environment came to be forced to adapt to man), the improvement was paid for by a radical increase in existential stress. The regions between the (imaginatively) expected and the encountered, or achieved, became vast. The fall of the mighty is hard: animals can make misjudgement, go hungry, or die, but only man can build castles in the air and find himself in an internment camp. Though not uniformly, existential stress so understood has been increasing throughout the history of man. We shall later see that the capacities of the individual mind to deal with the ever-increasing difference between imagination and reality reaches a limit when it finds itself up against the cooperative functioning of a community of minds. Then and there, we shall have to argue for the emergence of a new integrative level, that of the societal.

4   Between Life and Mind

Since the mind retains many features of its evolutionary development, we ought to be able to identify among our mental functions some that correspond to the transition between animal and man. If we assume that the position of the transitional apes was intermediate between the animal kingdom and emerging man, we also ought to assume that those portions of the mind which represent that transition will themselves have an intermediate position between life and mind. We already encountered an analogous situation when we sought to identify structures between inanimate and living matter.

One way of entering this issue is through the methods pioneered by Sigmund Freud. We shall consider the manifest content of dreams, those "royal roads to the unconscious." The predominant temporality of dreams is that of pure succession; although not everything happens at once, whatever does happen is informed of a curious fore-and-hind knowledge. Past, present, and future are, in a way, co-present; I know ahead of time what is going to happen, I proceed (in the dream world), and it happens. Connectivities among events are best described as those of magic causation; if encountered in the waking state, they would leave us dumbfounded. But in the Umwelt of dreams, the connections appear obvious and unproblematic. Often we are both the observers and the observed; the boundaries of the self, if at all, are very loosely defined. Dream Umwelts are ruled by some ancient paleologic, not unlike that of fairy tales.

Consider now the experiential basis of the widely employed metaphor that describes time as passing slowly or rapidly. In a recent study of the affective correlates of the experience of duration (our mental present) Peter Hartocollis argued that the feeling of the passing of time originates in anxiety that relates to need fulfillment. From clinical and theoretical work he concluded that when the tension that gives rise to the mental present derives from primitive feelings, such as usually arise from the id, then time is experienced as slow; if the tension derives from the dictates of the superego (roughly, our noetic Agent), then time is experienced as fast.* He writes that, in general, "the speed of time is determined by the prevailing source of the experienced tension." In childhood, wishes are strong and fundamental, time passes slowly; as we age our desires tend to become more abstract, socially conditioned, noetic; time seems to pass faster and faster.

It would seem that the "slower and slower" time of human experience corresponds to decreasing existential tension. The respective states of mind approach conditions that resemble the hazy creature present of advance animals, wherein everything beyond the immediate past and future vanishes into non-existence. The "faster and faster" time corresponds to an increasingly sharp separation of past from future, until, for the dying man, all his life appears as having gone by in an instant.

Putting the various arguments of this section together, we have evidence that the lower levels of the mind determine a temporal Umwelt which is more primitive than what we normally call mental, but too sophisticated to be biological. It is populated with imagery, albeit without a present, and without well defined identities. It is as though certain functions intermediate between those of the purely mental and purely biological were banished by

---

*Await the expected pain or sadness; wait for hours; look at the clock: only 20 minutes have passed, or perhaps only 4 minutes. Watch Ingmar Bergmann's "The Magic Flute," or share the company of your beloved; a whole world has passed by: look at the clock, it has been only 65 minutes. These are feelings from the id. – Start in the morning, begin and finish the working out of ideas that had clamored for recognition at 5:00 A.M. Hardly a dent had you made on the task, you just began warming up to it a sentence or two ago. Look at the clock. It has been six hours. These are feelings controlled by the superego.

the mind and the body, and sandwiched into a buffer zone between them.

The class of functions between those of the body and those of the mind may be identified with the human unconscious. That there are mental processes of which we are unaware might well be a necessity so that our bodies and our minds remain sufficiently distinct.

One of the many results of this evolutionary policy is the difficulty of regressive sharing: the difficulty of the imaginative sharing of lower temporal Umwelts. The mind classifies not only the biological functions of its body as those of the non-self, but also some of the lower mental functions. For instance, short term memory and recall can function without involving the feeling of personal identity. Playing a musical instrument may even feel as though performed by another person; the self can direct its attention elsewhere while the details of musical actions are being organized. "I" can still make myself become aware of what I am doing if "I" so desire because the task is sufficiently removed from the strategist. Working psychiatrists are continuously made aware of the many tricks whereby the mind represses and censors all information that might threaten its integrity. The self insists on keeping its distance, for only then can the mind act as though it were omnipotent. Since the collapse of the unresolvable conflict of the mind (to be discussed later) would spell an end to mental functions, it is that conflict which is being protected. Maintenance of the conflict secures for the ego its greatest assets: the separation of futurity from pastness, the clear distinction of the self from the non-self, and the privileges of free will.

When discussing the life cycle, we found that the soma regresses periodically to its genetic form so as to create new phenotypes. Likewise, individuals regress from conscious experience to the hazy domain of memory, expectation, and selflessness. During sleep, the impressions of the day are integrated according to the paleologic of racial experience. As the newborn, so the morning-man awakes for a new cycle of his life. Often, the regression of life to the genes is amiss and the newborn is ill or moribund; likewise, the nightly regression into dreams might not be enough to solve the burdens of the conscious self. In any case, it is in the unconscious levels of the mind that selection among autogenic images take place. Our conscious desires and needs are organized according to the dictates of ancient paleologic, and are prepared for presentation to our superego.

The horror visited upon man's mind if circumstances suggest loss of identity is well known from clinical practice, from politics, and from art. If man could convince himself that he is a predictable machine (that his highest Umwelt is the eotemporal, or even the biotemporal) then, wrote Dostoyevski in his *Notes from Underground,* he would contrive destruction and chaos to assert his personality. And if he were told that even these reactions could be calculated beforehand "man would purposely become a lunatic, in order to become devoid of reasons, and therefore, able to insist upon himself." Once the integrative level of human self was born and the nootemporal Umwelt emerged, its features came to be protected against collapse into the integrative level of life no less strongly than life protects itself against collapse into the inorganic.

5   Language

As far back as can be traced, people seem to have known that to communicate in language bestows fearful powers upon those who can do so. It is not surprising, therefore, that oral language shrouds the beginnings of language in mystery, and that the search for the origins of human language has great intellectual fascination.

Opinions as to what the study of language should include have varied greatly. At the end of the 18th century Herder insisted that language is the door to the soul; half a century later Wilhelm von Humboldt described it as an intellectual instinct of the mind. The most versatile modern approach is that of semiotics, the science of communication in terms of signs and signals.

Each integrative level of nature has its level-specific language and man, as a hierarchical system, is open to his environment on all of these levels. The human frame as a physical body is part of the eotemporal world, and hence its languages are those of chemistry and macroscopic physics. Appropriate languages on the macromolecular levels are, among others, genetics, virology, and microscopic physics; indistinguishable members of a set* form prototemporal aggregates, so they communicate through statistic and probability. It is only at the level of the mind that spoken tongue emerges; it is a class of articulate vocal sounds (and the earliest mode of communication that was described as "language"). Along the hierarchical ladder each higher "language" (used here in its extended meaning) subsumes the morphologies of those beneath. For this reason even abstract statements of formal communication tend to retain traces of their origins. As biosemioticists would quickly point out, to support such ideas we must understand them; to support weights, we must stand under them.

Microorganisms, sponges, fungi and the lower metazoan invertebrates are restricted to chemical and tactile communication. Interestingly enough, this alone is not a limitation on the development of a sophisticated language: it has been estimated that a chemical substance, modulated under ideal conditions, can generate 10,000 bits of information per second. Since this is far in excess of the capacity of human speech, there have to be some other reasons why evolution has selected verbal, and not chemical means of communication for Homo sapiens.

Let us recall the argument that the individual cannot normally see his total body, or explore it by touch, but his auditory system does offer certain immediacy about the self: I can hear my utterances in ways that compare well with hearing the utterances of other.** In spoken language the loop made of the nervous system, of our faculties of speech and of hearing offers a

---

*Gas molecules in a container, citizens on vacation, 50 lb. mailbags, workers in a ballbearing factory, 1977 model VW Rabbits.

**According to W.H. Thorpe (1961) this is also an important element in the phylogeny and ontogeny of bird songs.

self-referential intimacy second to none. I propose that it is because of this privileged position of voice that articulate vocalization was selected for the purpose of inserting, into the inner landscape of the mind, the symbol of the self. I have already suggested that articulate language and selfhood are likely to have arisen together; thus in a very literal sense, "in the beginning was the word."

Unlike chemical sensors, visual and auditory systems demand sophisticated multicellular transmitters and receptors, although the evolutionary packaging of hearing and sight are different. The complex neural networks necessary for seeing are mostly in the retina (some 5×10^6 neurons, forming what is sometimes referred to as an external brain). The neural system that processes audio signals is in the brain. Visual information may be encoded by spatial code, whereas auditory information demands temporal organization. If the self had to be defined by process description, as I argued that it had to, the temporal nature of the auditory system was there, so to speak, for the asking.

The carrier of the speech sound is produced by the vocal fold, in an adaptation of the cyclic order of life to the needs of the aging order, and to those of the mind. But information is mainly in the complex modulating patterns of the carrier frequencies, and not in the carrier frequencies themselves, not even in languages that inflect. Speech, therefore, involves a very wide spectrum of oscillations which demand motor action, as well as the appreciation of short and long time intervals. The pitch of a speech sound may be 5000 Hertz, with each cycle lasting for 2×10^{-4} sec. The single modulating envelope necessary to reproduce Richard Wagner's tetralogy of musical dramas, *The Ring of the Nibelung*, may be 2×10^{-5} Hertz, with each cycle lasting for 4×10^4 sec. It has been argued that much of the evolutionary enlargement of the human brain is the result of the need to develop a neural system that can handle this immensely broad range of sounds. Extending our earlier notions further, it is likely that language, the capacity for the fine structuring of time, long term expectation and memory, and selfhood form a single evolutionary syndrome.

Language makes possible the accumulation and the passing on of acquired knowledge and is intrinsically a communal affair. Private language makes sense to the extent that it forms a pool, somewhat larger than the store of conventional signals: as one tries to grant "to airy nothings local habitation and a name," one usually fishes in this pool for existing words, phrases, sounds, to be modified. However, it is difficult to imagine how human language could even have arisen were it not a collective enterprise. The embryos of many birds give clicking noises before they hatch. These clicks were shown to be essential for normal communication between mother birds and their young; as signals exchanged among the eggs, they synchronize hatching. Of what use would be the click from a lonely egg, or the dance of a single bee? Man does have voices that cry in the wilderness. But they can prepare "the way of the Lord" and help the species survive only if someone hears them. Still, the inner component of human language is bottomless. If we were to try to trace its origins very seriously, we would find ourselves descending along the biosemiotic ladder from man down to the prototemporal language of molecules.

During the last few years it has been demonstrated that chimpanzees can appreciate pictorial symbols and even answer questions that involve conditional statements. With ingenuity and effort, and through participatory methods, chimpanzees can surely be made to reach the upper limits of their genetic endowment, just as Americans reached the genetical limits of body height after being well fed for a few generations. E. O. Wilson, in his study of sociobiology, remarks that the problem with animal communication is that its messages cannot be manipulated to provide new classes of information. This brings to mind our earlier conjecture that human language arose as part of a new class of phenomena, when the creation of selfhood opened up an inner landscape of immense wealth and plasticity. Until chimpanzee brains can be made to cross that ill-defined boundary between the complex and the very complex, an unlikely feat except by genetic manipulation, a radical discontinuity, (but not at all a complete dichotomy) will remain between human and chimpanzee languages. Even the simplest of humans will live in a reality different from that of the best educated apes.

Consider now that children vocalize long before they read, then vocalize while they transfer their perceptual modality from the audio to the visual channels of the mind. The speech of the child which, in its early forms, reveals an inability to distinguish the self from the non-self (misleadingly called "egocentric speech") goes through the states of repetition, monologue, collective monologue, and only at the end of this gradual process of self-definition does he enter the phase of socialized speech. First it is for play, only later for communication. It is difficult to resist the idea that, in broad lines, the development of the language in the child, from perceptual system to communication, recapitulates the development of language in the cultural history of man.

Our pre-human ancestors surely knew how to utter certain sounds; slowly, so some arguments run, they learned to separate those which related to feeling from those that related to the early intellect. Shouts of pain or danger came to be distinguished from noises that related to tool or shelter. As it does in the child, so the gradual appearance of selfhood in man is likely to have become the paradigm of oneness; with one and the many, came the potentiality of quantifying experience. There are reasons to believe that although the brain is an analogue device, the invention of quantification made it possible to operate it in a digital mode; in its turn, digital operation seems to be a prerequisite for articulated speech. The quantification of experience was probably also necessary for the reliable transformation of sense experience into symbol. The external world could now be charted on to the inner landscape as onto a stage upon which a number of identities (friend, foe, food, and the self) played their deadly games. It also became conceivable to abstract invariances from feelings and other sensations whose sources could not be identified. All along this path, physical opportunities, of course, begot mental opportunities; thus the dexterity of the human body was surely necessary but not sufficient for the development of the brain and the mind. In any case, with a private world of meaningful experience the metaphor-making capacity could emerge. With it appeared a deathless world of symbols, pitted against the visceral knowledge of inevitable death. Our ancestors' lives came to be informed of an unresolvable conflict conferred upon the mind by the knowledge of time, one between individual death and deathless eternity. I wish to turn the argument around and propose that what we call mind is the source of behavioral manifestations that center on this unresolvable conflict. Perhaps nootemporality did, indeed, evolve so as to hide something that we surely know: the certainty of our individual passing.

I would envisage a syndrome of coemergent qualities: the definition of the self; expansion of delayed expectation to long term expectation; expansion of recall to long term memory; invention of language, and the discovery of individual death. These and related qualities are mutually reinforcing. Improved language skill increases the potential content of memory storage; this makes possible a richer harvest of autogenic imagery and with it an increased store of expectations. All these lead to increased constitutive tension because there now exists a larger store of expectations which may differ from conditions actually encountered. These steps would polarize futurity and pastness and assist in separating the nootemporal Umwelt from its biotemporal antecedent. Considering the several mutually reinforcing processes mentioned here, one would expect an explosive, rather than a gradual advance in the encephalization of early Homo. The strikingly rapid appearance of the humanoid features of the brain suggest that such an explosive, rather than a gradual, emergence of man's mind, was indeed the case.

VI Cosmic Images

IT COULD HARDLY BE BY CHANCE that the world, as understood in various epochs and at different places, has (1) always been judged to be best describable by the favored modes of knowledge, and (2) always been held to respond to the preferred means of acting upon the environment. Rather, this historical phenomenon exemplifies the power of the extended Umwelt principle. Whereas the Umwelts of animals are determined by their slowly evolving effectors and receptors, the worlds of man are determined primarily by the expanding capacities of his mind.

The great universal cosmologies of the ancient world included elements of ethical and aesthetic judgements, together with explanations about the earth, the sky, and society. But in the course of Western intellectual history, universal cosmologies gave way to a number of diverse and narrower disciplines: narrative cosmologies, history and geology, and scientific cosmology.

I shall argue that whether primarily of a scientific or of humanistic texture, the changing images of the world are complementary to the changing self-descriptions of man and society. Man and the universe remain boundary conditions to each other.

1 Timekeepers

The purpose of timekeepers is said to be that of measuring time. This function, contrary to popular opinion, is only secondarily a technical virtuosity. Primarily it is a metaphysical enterprise designed to coordinate the earthly organization of man's life with the temporal order of the universe.

The origins of archaic timekeeping are surely in the time-binding capacity of living matter. There is no period in the history of life when the cyclic processes of night and day and those of the seasons have not operated. The earth was spinning for at least one billion years before living matter appeared, and another 3 billion years before man emerged; how this cyclicity came to be recorded in living matter was the subject of earlier discussion. There are no periods in the known history of man when he did not express, through art and artifact, his desire to challenge the powers of passing time. That we describe the temporal organization of human acts in terms of suns, moons, and seasons (days, months, and years) and not vice versa, attests to the unquestioned authority of the rotating heavens as the primary clock and mover of all things.

It is not possible to identify the beginnings of the clockmaking trade. However, there exist many curious artifacts of certain uniform appearance and bearing systematic toolmarks, which, so some archeologists believe, were lunar calendars made perhaps 30,000 years ago. Early practices of timekeeping are indistinguishable from measurements and interpretations of the motions of stars and planets, performed in ways intelligible and useful to the maker of timepieces. Archaic astronomical knowledge was the basis of a complex symbolic system: deities and animals were projected upon the cyclic heavens perhaps as early as the 8th millenium B.C. and the famous Stonehenge in Wiltshire, England is believed to have served as a luni-solar calendar some 3,500 years ago.

The earliest known sunclocks are Egyptian shadow clocks, dating from the 10th to 8th century B.C. Extant stone sundials have been dated as early as the 4th century B.C.; their dials bore zodiacal markings which came to be replaced by engraved gridworks. Such gridworks attest to the ancient conviction that the shadows follow a reliable temporal routine: the sun does the moving according to the judgement of the heavens; the dial divisions reflect the life-style and needs of the dialer. In its turn, the geometrical nature of the gridwork suggests quantification by numbering and hence naming the divisions.

The sun is not the only astronomical body whose motion has been judged sufficiently reliable for the regulation of collective and individual practices. A plumb-line called *merkhet* is extant from Upper Egypt of 7th century B.C. It was used for determination of the hour of the night by observing the transit of selected stars across the meridian.

Hipparchus, astronomer and mathematician of 2nd century B.C. Alexandria, is credited with the invention of the astrolabe, a scientific instrument which in its medieval form has sometimes been described as the mathematical jewel. In twentieth century terms it is an analog computer that relates observational data on the apparent motions of stars, planets, and the sun, to the local time and geographical location of the observer. It links the cosmic affairs of the heavens with the earthly affairs of man through the temporal order that subsumes both.

There are reasons to believe that the origins of Western timekeepers may be found in certain advanced technologies of the first century B.C. employed for the task of modeling the motions of the heavens. Thus, the ancestors of Western clocks were early planetaria, and forerunners of what later became known as astronomical clocks. Of course, planetaria are not abstract statements about the heavens, nor are clocks models of human activities as though people were separate from the rest of the world. Clocks as planetaria and planetaria as clocks implicitly involve both man and the heavens. A dial type wristwatch is a Platonic device even if used at the Stock Exchange, because its hour hand models, (at twice the angular velocity), the rotation of the heavens, and is thereby the "moving image of eternity." Digital clocks are Aristotelian devices because they measure time by reference to the "number of motion in respect of 'before' and 'after.'" The before/after relationship is smuggled through the clockmaker's sense of time, a feature of his noetic Umwelt.

Time measurement necessarily involves at least two processes; for example, the oscillation of a pendulum and the crossing of the meridian by a star; the motion of the shadow of a stick and the cycles of my digestive system; or changes of the (imaginary) pointer in my calendar and the changing of the seasons. One of the processes is often hidden. For instance, if I assert that my wristwatch is accurate to within five seconds a day, I must be referring to a time signal, such as a beep I have heard on the radio. Or, I might claim that "It is just noon." The complete sentence may be "As I walked down from the third to the second floor, I heard the noon bell ring."

Let us call any set of instructions, or specifications on how the readings of one clock are to be translated into readings of another clock, a *time scale*. Thus, sidereal months transform to mean solar days through relations partly empirical, partly theoretical. A plot of the rhythm of human beard growth (16 days) is an instruction on how to change readings of beard-ounces/day to calendar readings. Cesium clock readings are judged accurate by comparison with the lawful components of the earth's rotation. Time scales are operational equivalents of the principle of the unity of time, since the mutual interchangeability of all clock readings can derive only from something that all timekeepers share. This "something" is the clockmaker's belief in the unchanging character of natural laws upon which the functions of the individual clocks are believed to be predicted. This holds for Egyptian shadow clocks as it does for atomic clocks.

Which particular clock is judged more accurate depends on the confidence in the lawfulness that is said to control the clock. Historically, clocks have always been defined as devices whose regularities were judged reliable by the standards of the epoch; an ideal clock is a device that conforms to what we think nature does most predictably. For Christian Huygens in 1673 cycloidal cheeks made the isochrony of the pendulum perfect. For the contemporary physicist, no less mortal than Christian Huygens, an ideal clock is defined as one to which he is led by arguments appropriate to his current understanding of the physical world. Whether or not a decision can be made between two clocks as to their accuracy, depends on whether or not there exists an acceptable theory that connects their readings; inaccuracies are usually assigned to the clock whose workings we do not understand. Thus, circadian rhythms of rat temperatures are instructions on how to change rat clock readings to laboratory clock readings, and vice versa. Since the rat clock is still very poorly understood, whereas the laboratory clock is not, the unpredictable variations in the comparisons are attributed to the rat. When an astronomical clock, such as the rotation of the earth, is compared with a hydrogen maser clock, the unpredictable differences in reading are attributed only slightly to our ignorance of the laws of atomic radiation and mostly to our ignorance of the laws that control the motion of the earth. That the Constitution of the United States of America was signed on September 17, 1787 is also an instruction. It tells us how to connect calendar readings with historical events. The a priori unpredictability of that date is assigned totally to our ignorance of what the laws of history are, or might be.

The history of time measurement demonstrates a dialectical progression between, on the one hand, the increasingly sophisticated methods of constructing timekeepers and, on the other hand, the increasingly abstract mental constructs that must be used to explain them. For instance, the accuracy of Christian Huygens' pendulum of 1658 could be tested against Kepler's laws of planetary motion which were regarded as exact, and were formulated some four decades earlier. Kepler's laws could have become suspect as incomplete when in 1845 the French astronomer Leverrier found an unexplained variation in the motion of the perihelion of Mercury, using the improved clocks of his day. The reliability of current atomic clocks is tested against certain first and second order effects predicted by relativity theory, formulated over half a century ago. What we observe is the reduction of successively more refined ideas about natural processes to working devices that confirm these ideas.

That astronomical timekeepers straddle the world of stars and of man is evident; but the same holds for timekeepers which seem to have no obvious direct link to the universe at large. For instance, the rate at which large bodies oscillate (balance wheels or folios, pendular clocks, water powered clocks) depends on the inertial manifestations of matter which, through Mach's principle, connect all local motion with the distant masses of the universe. Clocks controlled by interatomic forces (tuning forks, oscillating crystals, atomic radiation) respond to their gravitational and inertial environments. It is not necessary, however, to examine all conceivable clocks. General relativity theory guarantees that it is the cosmic distribution of matter that determines the rate at which a clock ticks, and this alone would make all clocks dependent on the universe at large. But the issue is really more subtle.

In 1922 Einstein wrote that

> in the present stage of development of theoretical physics these ideas [solid bodies and clocks] must still be employed as independent ideas; for we are still far from possessing such certain knowledge of theoretical principles as to give exact theoretical construction of solid bodies and clocks.

That is, although in the equations of relativity theory space and time do not play independent parts, we must still think about rods-as-distances and clocks-as-time as independent ideas. After fifty years of analytical work the early caution was replaced by a belief in the certainty of physical truth. Thus, from the massive *Gravitation* of Misner, Thorn, and Wheeler we learn that

one defines an "ideal" rod or clock to be one which measures proper length as given by $ds = (g_{\alpha\beta} dx^\alpha dx^\beta)^{1/2}$ or proper time as given by $d\tau = (-g_{\alpha\beta} dx^\alpha dx^\beta)^{1/2}$ (the kind of clock to which one was led by physical arguments ...). One must then determine the accuracy to which a given rod or clock is ideal under given circumstances by using the laws of physics to analyze its behavior.

The formulas that might look mysterious and hence endow the statement with authority, need not concern us. To those who have the time to learn about them, they tell in mathematical symbols precisely what the parenthetic phrase tells in verbal symbols, to wit, how the ticks of local time and inches of local length relate to the distribution of the masses in the universe, to the state of motion of the rod and clock, and to each other.

Placing "ideal" into quotes recognizes that the capacity of a clock can only approach but never reach the perfection which the formula represents. A less obvious but quite fundamental issue, however, goes unrecognized. Identifying the physical description of the universe as the final and sole arbiter of what an ideal clock is to be, is only the latest exemplification of the truth that the universe "has always been judged to be best describable by the favored modes of knowledge, and always been held to respond to the preferred means of acting upon the environment." *(Sopra)*

"The more perfect the instrument as the measurer of time," wrote Eddington in 1928, "the more completely does it conceal time's arrow." Indeed, increasingly more perfect clocks among physical processes lead us only down the path to the atemporal world of light. How do we get away from this necessary scientific provincialism? We should examine, I believe, increasingly more complex clocks instead of increasingly more ideal ones. Instead of oscillating atoms we should consider evolving genes, feeding fruit flies, and migrating elephants. With the privilege of having clocks that do say something about future, past, and present, we are going to lose reliability: the more complex the clock, the less assurance do we have that the next tick of oscillation will really take place.

With our broad understanding of the relationship of time-keepers to time, let us have a look at the modern version of that cosmos to which clocks and calendars relate the earthly organization of man, and see whether we can find it in the cosmologist himself.

2   Astral Geometry

There is no single crucial change or idea that could be held responsible for the birth of scientific cosmology from the matrix of universal cosmologies. But one may select by taste and judgement certain ideas that are useful in representing the change. One such idea, eminently suitable for our purposes is that of the shortest distance between two points in space.

Clement of Alexandria, an early Christian intellectual, wrote about Egyptian *harpendonaptai,* or "rope stretchers." Their task was to create straight lines, for land measurement and for astronomical determinations, by stretching ropes. That the shortest distance between two points is a straight line appears to be intuitively so obvious that its truth has become a dogma. Upon this and other, unquestioned truths was built the imposing structure of Euclidean geometry, so satisfying to the mind and useful to the artisan.

From harpendonaptai until Immanuel Kant the trend has been to separate the conceptual-logical from the experiential aspects of geometry. This practice was consistent with the separation of the manual and spiritual traditions which prevailed with few exceptions, through antiquity and the Middle Ages. The trend changed, however, with the increasingly significant symbiosis of artisan and savant in the Renaissance flowering of European technology. One formal expression of this rapprochement may be identified with the name of Carl Friedrich Gauss who, at the turn of the 18th century, applied his genius to the possibility of constructing a geometry whose validity may be experimentally tested. To describe this new geometry Gauss adopted the term "non-Euclidean," after having first called it by the prophetic name of "astral geometry." The name intended to imply that such a geometry becomes significant at astral distances.

In studying the properties of surfaces, Gauss made use of the notion of the geodesic, which is a line within a surface and the shortest distance between two points on that surface. The nature of a geodesic was found to depend on a property of the surface which he named its curvature. The concepts of geodesics and curvatures was later extended by George Riemann to manifolds of arbitrary dimensions. Thus, when the four-dimensional formulation of relativity theory made space-time the metaphysical receptacle of all events, the Gauss-Riemann multidimensional geometry was available as an analytical tool.* Eventually, though,

---

*In a plane the geodesic is a straight line, on a sphere a segment of a great circle, on a doughnut a variety of curves. For a plane the curvature is zero, for a sphere a constant number, for a doughnut a family of numbers. In each case, including multidimensional spaces, the curvature measures the extent to which the geometry of that space differs from that of Euclidean space, which is taken as having zero curvature. The four-space of special relativity theory, containing only electromagnetic radiation, is of zero curvature, or, as is sometimes called, flat. General relativity theory provided for the inclusion of energy and matter and showed that the new four-space, unlike that of the electromagnetic universe, is of a

the geometry of relativity theory turned out to be something unique; it is a differential astral geometry, better known as tensor calculus. In the four dimensional formulation of physics, the need for specifying shortest distances between points remained as important as it had been in classical geometry. But what for the Egyptian geometers were two points in space, for the geometers of space-time became two events specified by their positions in space and their instants of occurrence in time.

It follows, especially by hindsight, that the connection between two events, appropriate to space-time, has to be a motion of some sort rather than a distance. We found earlier that in their atemporal Umwelts photons provide instantaneous connections between what we recognize as events. Since nothing faster than instantaneity makes sense, light beams may be identified with the fundamental nul-geodesics of space-time. The shortest distance between two points in the old geometry was replaced by the fastest connection between two events in the new geometry. The ropestretchers were let go and light-beamers were hired who, instead of holding lines made of fibers, hold lasers, carry clocks, and will travel. Even the name "geometry" is inappropriate for this new scheme, for that word implies earth measurement. The geometry of post-relativistic physics is a new discipline, an astral geometry concerned with the physical universe at large and not at all with earthbound matters.

Euclidean geometry is constructed from points in space, from lines that connect those points, and from whatever one can do with points and lines to generate surfaces and spaces. Astral geometry is constructed from events*, from geodesics, that con-

---

non-zero curvature. In our understanding the appearance of non-zero curvatures is a corollary to the emergence of the proto- and eotemporal Umwelts from the atemporal world of electromagnetic radiation.

*Here "event" is taken in its unanalyzed ordinary meaning as anything that happens.

nect those events, and from whatever one can do with events and geodesics to generate multidimensional surfaces and spaces.

The universe of scientific cosmology is that of the symbols of astral geometry. The great confidence vested in this cosmology derives from the awesome power of mathematics which makes certain complex things appear to be simple and suggests, therefore, a complete identity between mathematical formalism and reality. In 1623, sounding as though speaking from Plato's *Timaeus*, Galileo wrote that the Book of Nature is written "in the language of mathematics, and its symbols are triangles, circles, and other geometrical figures ..." Reflecting this Platonic stance, the received teaching of physical cosmology is that its description of the world reports on a man-independent universe. I shall argue that this man-independence is not the case.

A convenient entry in this issue is Riemann's challenge in 1854 "that the propositions of geometry are not derivable from general concepts of quantity, but that those properties by which space is distinguished from other conceivable triply extended magnitudes can be gathered only from experience." In one fell swoop, our sensory apparatus became part of the means whereby geometrical truth is to be tested: a most unplatonic view. Riemann proceeded to distinguish our experience of extension (of space) from the metric representations of that space, with the difference between the two becoming important when "empirical determinations are extended beyond the limits of observation into the immeasurably great and the immeasurably small."

> That space is an unlimited, triply extended manifold is an assumption applied in every conception of the external world .... The unlimitedness of space has therefore a greater certainty, empirically, than any experience of the external. From this, however, follows in no wise its infiniteness, but on the contrary space would necessarily be finite, if one assumes that bodies are independent of situation and so ascribes to space a constant measure of curvature, provided this measure of curvature had any positive value however small.

With uncanny certainty, Riemann, then 28 years old, suspected that the Euclidean geometry of a priori truths does not hold for the very small and the very large.

In a playful way, let us imagine arc-like inhabitants of a circle traveling endlessly around it and finding it unbounded. Likewise, creatures living in the surface of a sphere would find the sphere unbounded. Yet arc-landers and sphere-landers, if they could develop geometries and study them, could infer from them a type of finity to their worlds; they could characterize these finities by such concepts as "area" or "volume" and describe their finite areas and volumes in terms of numbers, called radii or curvatures.

The prevailing opinion in physcial cosmology is that our four-dimensional world is unbounded but finite. Although in principle, if not in practice, we could travel in space endlessly, we can infer from astral geometry that the world is fi- ite. The finity of this unbounded universe of ours is characterized by such data as its finite age, the finite number of particles it contains, and finite volume (of a sort). A characteristic measure of this finity is the curvature of four-space, a quantity which, as explained above, measures the deviation of specific examples of astral geometry from Euclidean geometry. In the relativistic theory of gravitation the curvature of space is intrinsic to space-time and may be determined by measurements performed from within. In the words of Misner, Thorne, and Wheeler, the curvature must be "defined without any use of, and repelling every thought of, any embedding in any hypothetical higher-dimensional flat mani- fold." For, "geometry is a dynamic participant in physics, not some God-given perfection above the battles of matter and energy."*

Nicholas of Cusa, colorful Neoplatonic cardinal of 15th century Tirol, also asserted that the universe is boundless, by which he meant that it has neither circumference nor a center. "It

---

*Subsequently space-time does come to be embedded in "superspace [which] is the arena of geometrodynamics."

is not infinite, yet it cannot be conceived as finite, for there are no limits within which it is enclosed." A world so constituted would remain "unintelligible without God as its center and circumference." For in God we find "a centre which is with perfect precision equidistant from all points, for He alone is infinite equality."

Consider now that the four-dimensional universe of relativity theory, that unbounded and finite continuum, could have equally well been described as infinite but finite. This is *not* a trivial point, because words color our mental imagery and guide our attention. One need not credit Cusanus with prophetic foresight; he was not a relativist. Nor can we exclaim that there is nothing new under the sun; relativity theory was unpredictably new when it first appeared. What we must admit, however, is that certain contradictions that seem alien to the mind may become acceptable if formulated in the dialect of an epoch. For Cusanus, whose summum bonum was faith, finity-infinity was unacceptable without the idea of God; to the physical cosmologist whose summum bonum is the scientific intellect, identical conclusions would be unacceptable without astral geometry.

Both cases illustrate, however, that after a given discipline adopts certain metaphors (coincidence of the contraries, space-time curvature, expansion of the universe) it becomes comfortable with them and tends to forget their awkward implications. Certainly, we do not live in the center of the earth; the earth is not in the center of our galaxy; our galaxy is not at the center of its galactic cluster. But, in physical cosmology we, and everyone else everywhere, are always in the center of the world. Some very radical steps have surely been taken in constructing such a cosmic image. Perhaps man, pushed out the door by scientific cosmologies, smuggled himself back, through the window.

The desire to embrace all that is knowable is already demonstrated by the Rhind Papyrus, a venerable document in the history of mathematics, written about 1700 B.C. It states its own subject as "rules of enquiring into nature, and for knowing all that exists [every] mystery ... every secret." Although no explicit claim is likely to be found that modern cosmology wishes to incorporate knowledge of "all that exists," this belief is nevertheless implicit in the literature.

The unstated belief that the physical universe subsumes man is suggested by the fact that the observer is assumed to be passive and is taken for granted: I was unable to find a professional work on physical cosmology that would include considerations of man in a non-trivial fashion. Of course he is there, quoted in mottoes, but his presence is like that of a bouquet of flowers on a new tractor: pretty but unnecessary.

Physical cosmology is an offspring of astronomy, married to physics and astral geometry. It inherited the pragmatic ontology of the hard sciences which understand truth as being "there" to be discovered. But an authentic study of the universe cannot afford to neglect man: I am positioned within that unbounded but finite object which is the world of scientific cosmology. It certainly seems to include me with skin, bones, and mind. (Or, does it?) Although it would be difficult to find a cosmologist who would insist that all of biology, psychology, sociology, and physics may eventually be derivable from Einstein's field equations, that tacit assumption has been part and parcel of physical cosmology and will remain so unless it is explicitly and categorically disclaimed.

Consider now the hierarchical structure of time. With very few exceptions, such as the works of G.J. Whitrow and P.C.W. Davies, the time of modern cosmology is identified with the same ongoing thrust which is implicit in human time, or nootemporality. The following is from a recent definitive work on *Gravitation* by Misner, et al.

> The aging of the human body is governed by the same electro-magnetic and quantum mechanical laws as govern the periodicities and level transitions in atoms and molecules. Consequently, aging, like atomic processes, is tied to proper time as governed by the metric [of general relativistic geometry] – though of course, it is also tied to other things, such as cigarette smoking.

I assume that in its lighthearted brevity "cigarette smoking" stands for all things peculiar to life in general and to human life in particular, such as the intellectual tension attendant to writing cosmologies. But, alas, there is nothing in the metric that would tell us what to mean by "now" or by "future-past-present," hence how can it govern aging?

Of course, those function of our bodies that are appropriate to the physical Umwelts must be included in the purview of physical cosmology. The lawfulness of light that reaches our eyes, that of heat which reaches our skins, the biochemistry of living matter, or the mechanical laws that restrain our motions are determined by the same family of laws that govern the totality of the physical universe. But the now-ness of living matter or our inevitable progress toward death are only restricted, but not determined by the rules of quantum mechanics or relativity theory.

The physical universe, including its temporalities, may be given articulate expression in symbols and signs appropriate to the nootemporal integrative level, because the language of the mind subsumes the lower languages, in addition to determining a region of rules of its own. The mind can render a self-consistent and authoritative account of the physical world through such means as astral geometry because, as I insisted earlier, the brain is more complex than the physical universe.

3   The Old Man with the Clock

What does the scientific account of the physical universe comprise, and how do people go about determining these features?

The physical universe is identified by its average density, its curvature, and its "age" (a rather tricky concept). The determination of these quantities is a complicated process. Their empirical bases are in observational astronomy and astrophysics. Each variable has its own problems of techniques, theorems, and assumptions. The cosmological portion of the calculations involves further theoretical work and further assumptions about the detailed structure as well as about the large-scale features of the universe. The final result is a mathematical model of the world, that is, a structure of mathematical physics whose behavior seems to agree with the observed behavior of the astronomical universe.

There are several equally plausible models available for a universe of astral geometry. All those that look promising permit the labeling of the large-scale conditions of the universe in terms of a single function (called the radius of the universe) of a single variable: time. They all suggest that the universe is evolving in a type of time which may be called "cosmological common." This is a universal, cosmic time that ticks away throughout the world, and resembles Newtonian time in its commonality to all galaxies (which are the natural units of matter on the cosmic scale).* With respect to each instant of cosmological common time, the average curvature of three-space is taken to be the same at every point within the unbounded but finite universe. Thus, very significantly, the curvature itself is a function of cosmological common time, shared by all observers. At the singularity, which is a mathematical designation of the first instance of the temporal world (and has been better known as the instant of Creation) the curvature is assumed to have been infinite. In our epoch, which is some 10^{10} years after the singularity, the curvature is ≈ 4.75. Eddington once remarked that when we perceive that a region contains matter we are recognizing the intrinsic curvature of the world. Yet the curvature is much more than a measure of local matter density; its evolutionary decrease signifies the thinning out of a vast but finite mass in a vast, expanding, but finite, (though unbounded), universe.

The expansion of the universe is a difficult but clear concept. Its clarity derives from the Platonic beauty of mathemati-

---

*There are many ways one may relate processes in time to time; the cosmological-common time is one of them. Local variations from cosmological-common time are not due to cosmological features but to the motion of observers (real or imagined) with respect to local galactic frames. Other relationships between time and process are known as constitutive time, relational time, absolute (Newtonian), absolute (non-relativistic), special relativistic, and general relativistic.

cal physics; its difficulty from the fact that those who try to explain it usually take the metaphor of "expansion" for real.

Take four balls, name them galaxies. Construct a system where they are mutually equidistant. The arrangement that satisfies this demand is a tetrahedron with a galaxy at each corner.

Now try to construct a system of five mutually equidistant galaxies. Use your yardstick to measure distances. The task is impossible.

Extend your system to astral dimensions and use the only available yardstick: light. To calculate the positions of your galaxies use the instructions of relativity theory. You shall be able to construct a five-cornered system of equidistant galaxies, but your space will cease to be Euclidean. The space of such a system corresponds to that of the actual universe.

As we learned earlier, a useful measure of the topology of the new space is its curvature; it tells you to what extent the geometry of the new space differs from the geometry of the Euclidean space.

Start with a space of infinite curvature – well, of a very large one. All space there is, is within the boundaries of your space. No physical meaning can be attached to anything external to it. Let the curvature decrease, approaching asymptotically the flat Euclidean space of zero curvature. Your universe is now expanding.

As the universe expands or, if you wish, flattens out, a number of other things might also be happening to it. There were cosmological theories of continuous creation afoot for a few decades, suggesting that as the universe expanes, matter is created ex nihilo, (but not by God), to fill the increasing distances among galaxies. Apart from its curious metaphysical difficulties, this theory appears to be empirically untenable. But the unease which suggested it is a legitimate one: what else, besides a geometrical expansion or flattening takes place? I think the answer is this: we are flattening out or expanding, (pick your metaphor) into the future. New matter need not be created ex nihilo but novelty *is* continuously created. The changing geometrical curvature of the world is a function of cosmic time. So the physical corollary of the geometrical change is the aging of the universe.

But alas, I already argued that the highest temporality of the physical world is one of pure succession in agreement with Davies' conclusion that the only type of model of the universe that is in good agreement with observation employs the "exotic possibility of topologically closed time."

Presently we shall try to understand what may possibly be meant by saying that the universe flattens out, expands, or ages into novelty. Beginning with nootemporality, we shall imagine ourselves journeying from our own epoch back toward creation. We shall always use whatever clocks may be appropriate for the different portions of the journey, while also remembering that back home, in our nootemporal world, time is measured in years, months, seconds.* The journey will remind us of the famous Cheshire Cat in *Alice in Wonderland,* because it knew how to vanish "quite slowly, beginning with the end of the tail, and ending with the grin, which remained some time after the rest of it had gone."

As we go back a mere few million years the nootemporal Umwelt of man vanishes: there are no kitchen clocks, no lunar calendars, and no individual deaths. In another 3 billion years life itself vanishes and with it, the possibility of meaningfully defining a present; the world becomes one of pure succession. The average temperature of the universe is still only a few degrees Kelvin, astronomical processes go on, but time has lost its arrow. Yet undaunted, we keep on traveling. In another billion years the earth, then all the galaxies, vanish, and with them the possibility of using eotemporal processes as clocks. The temperature is perhaps 1000 °K, slowly increasing. Our clocks from now on must be prototemporal, that is, we cannot identify when precisely something happens. Also, events and things often become indistinguishable. But nothing can stop the time traveler, as long as he only imagines the travel. Millenia go by and temperature rises; at about 10^4 °K the universe changes from a matter dominated to a radiation dominated state. Day and night ... and the last (that is first) day is coming up. In some 12 hours the primary elements and their isotopes vanish; there can be no more prototemporal clocks. The temperature rises to 10^{12} °K; a cubic cm. of the world

---

*Cosmological common time is sometimes measured in terms of the expansion factor, a dimensionless quantity, or in units of red shift. However, since these "roses by whatever names" bear precise functional relations to proper time, their uses amount to euphemisms, as far as the fundamental issues of time are concerned.

has a mass of 10^{13} grams or ten million metric tons. Soon we are only about 10^{-24} sec. away from the singularity of creation. But this amount of time, as we learned earlier, is the atomic chronon. Now even the grin of the Cheshire Cat must fade away because by the general relativistic red shift, in this atemporal Umwelt, where matter is of infinite density, our last conceivable clock has stopped.

Since an atemporal world cannot support connectivities, not even probabilistic ones, we might as well regard the beginning of the universe as a purely contingent event. We might also agree with *Timaeus:* "but the principles which are prior to these only God knows, and he of man who is the friend of God."

The beginning of cosmic time, not unlike the beginnings of our personal memories, vanishes in an atemporal Umwelt. We have not only thought-traveled back to the beginning of the world, but also out of it. Placing the beginning of the world of astral geometry at the epoch of the Big Bang 5×10^{17} seconds ago is as misleading as Bishop Ussher's famous date for the creation of the world, Sunday, October 23, 4004 B.C. Both estimates are justified by the authority of the prevailing mode of knowledge. Both extend the validity of nootemporality to all integrative levels, and differ only by a scale factor of 10^6. In the hierarchical scheme of time one must insist that the age of the world be understood as a development of temporalities. It would follow that it is this development which we perceive as the flattening out, or expanding, of the physical universe.

It has been argued convincingly that geometry be regarded as a branch of physics because geometric laws, as they are understood in astral geometry, are subject to empirical test. But if geometry is a branch of physics now, then it must have always been so whether realized or not.* If the laws of geometry as

---

*Geometries may be discussed as mathematical propositions whose validity is determined entirely be self-consistency of postulates and theorems, with no overt reference to the world of the senses. However, as we shall see, logic itself is of evolutionary origin, having its source in the selective advantages it offers to the species, hence mathematics is also empirical in its roots.

understood today may be refined and corrected by the extended senses (such as telescopes), then the postulates of Euclid, formulated some 2300 years ago, must have been obtained through the unextended senses. Our knowledge of how postulates of mathematics derive from sense impressions is very poor; but short of pure idealism, one must assume that they do. But then it follows that the validity and limits of mathematical truths must be interpreted in a broad framework, with the assistance of all knowledge that helps reveal the processes which tie sense impression to mathematical law: perhaps psychology, biology, sociology, perhaps linguistics.

Consider again that the space of astral geometry is not embedded in a higher order space: it is bounded (that is, self-contained) and infinite. But its large-scale conditions do change as a function of a single variable which has a present value and a uniform applicability to the whole universe (for the relativistic oddities are but local deviations). The same variable also has a beginning at the singularity. By the principle of parsimony we must conclude that that variable is noetic time.

Even in the description of the universe through astral geometry, nootemporality thus functions as the organizing principle of our perceptions and thoughts. It permits the mind to render an articulate description in symbols and signs of the structure and functions of the physical world, whose Umwelts are common with the lower Umwelts of the psychobiological organization of man.

Cosmological common time, then, is Kantian in that it is an a priori form of perception; but it is also Darwinian, for it is a mode of perception which has evolved through time. This curious stance, consistent with the extended Umwelt principle, suggests that what we recognize as the evolution of the cosmos in time, from past, to present, to future, is not "there" to be discovered but is a representation appropriate to the nootemporal Umwelt of the mind. By the hierarchical structure of integrative levels, the physical world is such that it permits such an imposition (for, as I shall argue later, the determinants of the nootemporal world are left unrecognized in the languages of the physical universe).

The cosmic image of astral geometry, though it exudes an atmosphere of truth independent of man, is but the universe of a sufficiently old man with a clock. He has memories and expectations, he integrates within his identity a hierarchy of Umwelts, he bears his unresolvable conflicts, and he insists that he himself witnessed Creation "when the morning stars sang together."

4 His Home and his Story

The old man was not always as meticulously hidden as he is in scientific cosmologies.

Following the Pythagoreans, Plato held that the preeminent characteristic of the world was its ordered beauty. The soul of man, being rationally ordered, was capable of becoming the world in miniature. The Greek concept of *Kosmos* applied to a living thing, namely, to a well ordered society, as well as to the orderliness of the world at large. Thus, the history of Western thought, ideas of social and political order, and of an ordered world external to society, were born simultaneously.

Plato's dialogue on cosmology, *Timaeus*, contains not simply a story of how the world was fashioned, but also a good helping of ethical and aesthetic judgement, and accomplished poesy. The universe, for instance, must be comprehensible, and must be the likeness of what is "always real and has no becoming," that is, alike to whatever is permanent. The intellectual ancestry of physical cosmology may be identified in the words of Timaeus with uncanny certainty. Plato's ideal of the absolute form is embedded in the metaphors of astral geometry; the prostituted images of these forms are the clocks and rods, never ideal.

The universe, Plato says, is created of two classes: "One, which we assumed, was a pattern intelligible and always the same, and the second was only the imitation of the pattern, generated and visible." Then, as we learn from the *Laws*, to which the *Timaeus* is a necessary introduction, the chief business of the state is to mold the character of its citizens so that they may assimilate to the vision of ideal political order, itself an image of the ordered beauty of the Greek universe.

Belief in the intrinsic mutuality of cosmic and social views was not limited to the Middle East and Europe. The perennial philosophy of China, in the words of Joseph Needham, has been "an organic theory of the universe which included nature and man, church and state, and all things past, present, and to come." Astronomical systems and calendars, complete with mathematical techniques, were the Emperor's ritual paraphenalia since time immemorial. The emperor had to know Nature's Tao so that the political order of the state might remain consistent with it. In Nathan Sivin's phrase, (1969), the Chinese theories of natural order and political order were "resonating systems, with the ruler as a sort of vibrating dipole between them ...." The throne of the Chinese Emperor always faced south, for the Emperor, as the pole star above him, was fixed in the center of the world, unmovable and eternal, directing harmoniously the affairs of the state while, in principle, doing nothing.

Not all universal cosmologies are as elegant as Plato's, or as practical as those of the Chinese; but they all incorporate into their teachings the collective of man. Thus, in the Babylonian *Enuma Elish*, written perhaps a thousand years before Plato, the Lord trods upon "the hinder part of Tiamat," Dragon of the Chaos, splits her skull, creates from it earth and heaven and delivers a sermon upon the calendrical divisions of the year. In Hesiod's powerful and often amusing *Theogony*, out of the void came Earth, Eros, and Darkness; and out of the Night came Light and Day; in due course came Law and Memory and a family of gods.* In Greek mythology Zeus himself bridges the gap between cosmic order and social disorder, acting as a mediator between the heavenly landscape of eternal harmony and the conflicting drives of mortal man.

Godhead and man met in the territories, temples, and cities of the Mesopotamia of the 4th millenium B.C.; they were that

---

*As we learned early in this book, Eros grew wings, became rather handsome, and liked to watch Aphrodite play five stones with Pan.

It is the necessary purpose and aim of the mathematician to show forth all the appearances of the heavens as products of regular and circular motions.

Ptolemy, *The Almagest*

The movements of the heavens are nothing except a certain everlasting polyphony (intelligible, not audible) with dissonant tunings.

Kepler, *The Harmonies of the World*

[Einstein's theories] dethroned spacetime from a post of preordained perfection high above the battles of matter and energy and marked it as a new dynamic entity participating in this combat. [They also recognized that] the proper arena for the Einstein dynamics of geometry is not spacetime, but superspace.

Misner et al., *Gravitation*

He was a philosopher, if you know what that was. "A man who dreams of fewer things than there are in heaven and earth" said the Savage promptly.

Aldous Huxley, *Brave New World*

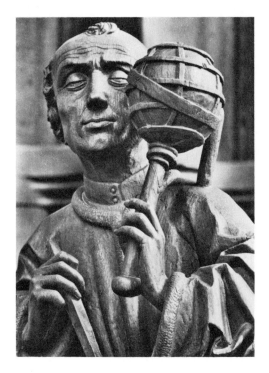

Ptolemy. Wood carving from the choir-stalls of the Cathedral of Ulm, made between 1469 and 1471 by Jörg Syrlin the Elder. Courtesy, Deutscher Kunstverlag, München.

THE CHANGING IMAGES of the world are complementary to the changing self-descriptions of man. Society and the universe are boundary conditions to each other, connected through the Old Man with the Clock. As though on a Möbius strip, the outer world of the cosmologist turns into his inner world, and vice versa, if we but dare inquire in sufficient depth.

many celestial archetypes, centers and navals of the world. In St. Augustine's 4th century city on earth, judgments about man's affairs on earth without reference to the affairs of the world at large were totally unthinkable.

The Mosaic cosmology of *Genesis* reveals all things along the step ladder of values; these values represent the judgment of God. We have light and heaven, earth and grass, days, nights, seasons and years, birds and bees. Finally comes man alone and in groups, with the capacity to disobey God's unchanging law and be ready for the Fall. From the story of the Fall follow the essential parts of the Christian world-view which delineate man's position in the world vis-a-vis the skies and the divinity. Herein the faithful finds the rationale of Christian soteriology and can learn those aspects of theology that are to guide his conduct.

But gradually, through the confluence of many rivers of thought the mixing of many modes of aspirations, and influenced by numerous and complex needs, Western methods of inquiry came to regard place and story as two distinct concerns. Geology, in its common, garden variety form, addressed itself to issues about that large body of water and soil upon which man lived. History came to deal with answers to this question: what happened on earth before I was born? These two departments of knowledge, though remaining distinct, came together in the Darwinian revolution. Geology gave history its final scale; evolution through natural selection gave the story of man an absence of purpose. And history gave itself a headache by not knowing whether it should primarily concern itself with the regularities that seem to inform the past, or else with the unique events in the story of man.

Historicity, or, rather, historical attitude, has its noble ancestry in the linear view of time that was born as the Hebrew Heilsgeschichte, or salvation history. In the words of S. G. F. Brandon (1965), salvation history is "the record of Yahweh's original deliverance of Israel and of his continuing providence according to [that] nation's deserts." Expressed differently, history as a development of mankind's story became conceivable when the direct action of God upon the affairs of man was first interpreted as a promise of deliverance. Henceforth, the path of Israel would lead from an unbearable present, through a series of victories against oppressors and enemies, toward a repose in the final Promised Land.*

The heavenly road building began with the vision of Isaiah who heard "the voice of Him that crieth in the wilderness, Prepare ye the way of the Lord, make straight in the desert a highway for our God." It was elaborated by Christianity into its own salvation history wherein the life and resurrection of Christ is the pivotal point of an otherwise evenly flowing time. At the end of this time we shall all be one in the land of green pastures and shining rivers where, finally, "sheep may safely graze." This writer hopes that the music played will be that of Johann Sebastian Bach, and the company we keep will be that of all those we have ever loved; and also that explanations of passion and knowledge will not be needed, for everything will be obvious, necessary, and free.

Beginning with the scientific revolution, the idea of Heils-geschichte was secularized into the ideal of progress, politicized into the dogma of unlimited benefits, and finally prostituted into slogans of the market place. "Progress" we hear from the public relations department of industry "is our most important product."

As already mentioned, from among the many children of universal cosmologies, history deals mainly with story, geology mainly with place, (the earth, that is) though neither is totally independent of the other. The two are joined together, and to other fields of knowledge, through the principle of uniformity of nature. Uniformitarianism is a rediscovery of a sort, a result of the great geological debates of the 18th century. In its simplest and

---

*See, however, Voegelin (1974), who argues that the unilinear construction of history, from a divine and cosmic origin, is a symbolic form of order developed by the end of the third millenium B.C. in the empires of the Ancient Near East. Voegelin describes this creative process as "historiogenesis." He sees historiogenesis as slowly giving rise to the genesis of world history during the epochs from the rise of the Persian to the fall of the Roman Empire.

most general form it states that the major epistemic features of nature (lawfulness, causation, intelligibility) are common to all times and all places. Explicitly or implicitly, this idea of the communality of all there is, must be extended to organic evolution as well as to history; assuming the opposite would lead to an unmanageable tangle of ideas.

But we learned that lawfulness and causation are hierarchically ordered and I shall argue later that so is intelligibility. Hence, the uniformity of nature must itself be understood as the uniformity of this hierarchy and its independence of time and place. However, insofar as temporality itself has a history of development, corresponding as it does to the consecutive comings about of new integrative levels, the claim for uniformity is not a claim for the actual existence of level-specific languages, causations, etc. before the integrative levels have come into being. When uniformitarianism is so understood, certain epistemic hurdles of cosmology, geology, and history may be resolved and these modes of knowledge arranged so as to please the old man with the clock. He likes to know his past, needs to be familiar with the furnishings of his home, and would like to know his future. So we will sketch some ideas about the relative explanatory and predictive powers of astronomy, geology, biology, history and sociology – derivatives one and all of universal cosmologies of tribal chiefs, prophets, and philosophers.

Astronomy is the oldest of sciences; in its modern form it is an exact science and it can make predictions of great accuracy. Its data comes almost entirely from information carried by electromagnetic waves so astronomy, exclusive of space exploration, has a built-in filter to keep minutiae out.

Geology is a very new science. It has been successful in explanation but not in prediction and the concept of precision does not seem to fit it, even though mathematical methods are increasingly useful both in physical and in historical geology. The material of concern to geology bears directly upon our senses: size, color, weight, smell. When we behold the earth, there is no "built-in filter." The advance of satellite-based geology promises to lead to the understanding of large-scale variations in surface structures and, through intricate correlations, to an improvement of our knowledge of physical as well as historical geology. Further refinement of our knowledge of the earth is likely to come from other sciences, such as minerology and paleontology. But, one cannot entertain the hope of being able to make accurate and long range predictions, not even in such matters of concern as earthquakes, volcanic activity, or weather.

New epistemic elements enter as we shift our attention from the earth as a rich heavenly body, to the earth as supporter of life, and then to life itself.

We have already considered the large degree of unpredictability immanent in the nature of life. Later we shall stress the fact that the language of each integrative level leaves certain regions of its own level unrecognized and therefore, undetermined. These regions gain in scope as we advance from the atemporal to the sociotemporal. When one reaches history, which is an aspect of the sociotemporal integrative level, the heterogeneity of lower order languages, causations, and undeterminacies makes the search for precise laws of history a senseless task.

When critical historians, or sociologists of positivistic persuasion seek to identify the regularities of the body social in ways that resemble the laws of exact science, they are employing the principle of uniformity in nature, but they neglect the hierarchical organization of that uniformity. By the arguments given, such neglect may be a fatal mistake: since the regularities of society cannot exist on integrative levels below the sociotemporal, using a language appropriate only to the lower integrative levels cannot yield meaningful results.

The multiplicity of lower-order languages remains a hierarchical unity because it is contained in the mind of the old man with the clock. Historical law is born when that mind thinks of it. Should the ancient pharaohs have been ever so brilliant, a Marxist interpretation of history, or Adam Smith's *The Wealth of Nations* would have appeared to them as utter gibberish; that to us they make sense has to do with our peculiar appreciation of the past.

As the universal cosmologies broke up into scientific cosmology, geology, and history, and as our growing capacity for different types of inquiries developed, the world came to assume aspects of apparently unrelated multiplicities. But the mind of man demands a unity in that multiplicity, and this has led to an unprecedented crisis in world-views. We shall return to this issue later in different contexts; presently, I want to sketch the background of the crisis in terms of the theme of this chapter, which is that of "cosmic images."

The self of the Old Man, it will be recalled, is a symbol of the internal landscape of his mind, distinguished from all other symbols in that landscape in the following way: there is nothing in the external landscape which would totally correspond to the self, yet it partakes in the goings-on of the external landscape as though it were an object totally out there. The self, in its course of ontogenic and phylogenic development, becomes defined and refined through continuous skirmishes with food, foe, friend, and other "selves." Analogous arguments hold for the collective self.

The collective perception of reality also contains many symbolic continuities. One of these, again, is partly external and partly internal to the collective itself, yet it is assumed to operate as do other continuities, on a stage of external reality. This symbol is the collective identity of nation, tribe, ethnic group or class, traditionally expressed in mythic visions. And there are good reasons why they are so expressed. For, consider that not even a creed freely adapted can make a group of humans totally identical. There is no single and foolproof test whereby all Americans, all Christians, or all boy scouts may be identified; witness the unending hassle of courts called upon to decide who and what belongs where. To be an American or a candlestick maker signifies "my" opinion that "I" belong to that group. The collective is imagined to operate upon a stage of other nations and other professions; but if I am a member of either group, I am never in a position to see that group in its totality as non-members see it. As does the individual self, so the group identity also becomes defined and refined through continuous comparisons with other collective selves. "One nation, indivisible, under God" makes sense only if there exist other nations, and if some of them, presumably, are not under God.

As does the individual mind, so the collective self also produces autogenic images. Some of these might correspond to external reality (another tribe, for instance); some may, but need not, have any external correspondence (such as the goals of a five-year plan); again others might not correspond to anything even in principle (such as the millenial expectations of utopian communities). Cosmologies are also autogenic images of the body social; they are also tested against individual imagery of the group members, and against cosmologies of other groups.

For various reasons, the epoch of the fall of universal cosmologies is also that of a radical increase in the interdependence of larger and larger collectives of man. Which offspring of the universal cosmologies is most likely to be acceptable to this emerging family of man? It is surely physical cosmology, because it deals with those aspects of the world in which individuals and nations do not differ: rates of free fall, average densities of matter, the frequency of the yellow sodium line, and the like. Indeed, scientific cosmologies deal with the lower, common Umwelts of stones, animals, and man.

But the cosmology built on astral geometry is void of anything that concerns the affairs of man as an individual, or in collective. Furthermore, the abstract character of the new cosmologies makes them unintelligible to all but a small group of people. Meanwhile, demands for normative statements about beginnings and endings, and about all the time in between, remain as strong as in past ages. Narrative cosmologies could easily speculate about beginnings and endings because their teachings were expressed in terms of human values. But issues of human concern are banished from scientific cosmology, and concentrated in the meaning of a single, unexamined variable, innocuously named t or τ. With the decline of cosmologies that spoke about the world and of man, our understanding of the universe was left empty of human significance.

There are also other problems. To remain consistent with the method of experimental knowledge, findings about our physical universe should be tested against independent observations of other universes, otherwise scientific cosmology may easily deteriorate into dogma. But by definition there can be no others, even if physical cosmologists talk about an ensemble of universes. Thus, uniqueness bedevils the fundamental issues of physical cosmology.

Uniqueness also haunts all inquiries into the affairs of the family of man, if that family is taken to include all people on earth. Comparative perspectives cease to be valid for a world-wide society. We shall return to this issue later, and shall assume the emergence of a world community to be necessary if advanced human civilization is to survive. But the perennial condition of man has been that of belonging to a community, pitted against other communities, as well as against the totality of the living and inanimate environment. It is this multiplicity that has permitted social development through methods of natural selection. For a world community, however, the definition of a symbolic continuity, "the family of man," in terms of comparisons with similar, symbolic continuities, is impossible. Unlike for tribes whose imagined destinies could be tested against the destinies of other tribes, there is no other humanity against which the collective self of all man could be defined.

What we do have, however, is a boundary condition that separates as well as unites the two universes: that of society and that of the physical world. This link is the mind of the same old man with the same old clock.

It seems, then, that in twenty-four centuries the Platonic mutuality of cosmos and society has not changed. Even in its most advanced form, the universe and the family of man are defined one with respect to the other. But the mind of man has certain freedoms which his perennial clock cannot command. In a very true sense, then, and in ways more frightening than Plato or Augustine would have imagined, we are alone: not as a true center of, but rather as a counterpoint to a universe of astral geometry.

VII Time's Rites of Passage

SO FAR we have been concerned mainly with the major integrative levels of nature; we have discussed the development of their temporalities, causations, unresolvable conflicts, and other determinants of their level-specific Umwelts. We did note, passim, that there exist certain processes in nature that are difficult to classify as appropriate for one or for another integrative level. They assume ambiguous positions between adjacent Umwelts: sometimes they may be classed with a specific integrative level, sometimes with the level beneath it. Such processes form the class of interfaces. Developmentally they seem to be short-lived. One knowledgeable critic of the theory of time as conflict found this feature important enough to remark that "nature does not make jumps, they used to say, but if you look at the hierarchy of complexity, she seems not to linger at intermediate stages."

Some of these "unpopular" states have already received our attention. We have referred to them in attempting to elucidate the functions of the integrative levels. Thus, we dealt with the unconscious, the interface between the biotemporal and the nootemporal worlds, in the context of the mind; biogenesis, a process that takes place at the upper regions of the eotemporal and connects it with the biotemporal, was discussed in the chapter on life; transitions in the physical world between adjacent integrative levels were discussed in chapters two and three.

In this chapter I wish to focus on the interfaces as a class. I shall argue that they have certain qualities that distinguish them from the less elusive adjacent integrative levels. Although the appearance of the unpredictably new is not at all limited to the interfaces, it is nevertheless only from these regions that new temporalities seem to emerge. Their marvelous alchemy comprises time's rites of passage.

We shall examine what appear to be policies common to all interfaces, up through that which separates the biotemporal from the nootemporal Umwelts. Then, I shall assume that the interface between the Umwelt of man as an individual and the Umwelt of society as a coherent unit is bound to exhibit the same regularities as do lower order interfaces. Based on this assumption, we shall seek an understanding of that unstable relationship between personal and communal identities that characterizes our own epoch.

1   Policies Common to the Interfaces

We shall consider four interfaces (atemporal/prototemporal, prototemporal/eotemporal, eotemporal/biotemporal, biotemporal/nootemporal) under five policy headings. Thus, we shall encounter each interface five times.

(a)   Metastability

As already mentioned, functions that we class as belonging to the interfaces are, figuratively speaking, disowned by both of their adjacent integrative levels. With a term borrowed from chemistry, where it has a number of closely related meanings, I shall describe this condition as *metastable.* *

O In chapter two we concluded that in the physical world we are dealing with three interacting but nevertheless distinct integrative levels: that of particles of zero restmass, that of particles with non-zero mass, and that of bulk matter. The many examples of overlapping do not represent conditions intermediate between the adjacent Umwelts. Thus, we have suitable methods to deal with waves and other methods to deal with particles; in many cases either of the methods may be used and we are free to think of certain functions as either those of particles or those of waves. But I know of no physical state that would correspond to energy

---

*"Marked only by a slight margin of stability" and "a state of equilibrium in which change to more stable forms does not take place spontaneously." (Oxford English Dictionary).

between the radiative and particulate forms. One can explain the famous double-slit experiment in terms of light as photons, light as waves, and light as both, but not light as between waves and photons. The creation and destruction of particles are assigned zero duration, which arises naturally in perturbation theory in which transitions are not analyzed. But I should like to speculate that intermediate forms may nevertheless exist, that is, the transitional times would be small but finite if we knew how to do the calculation. If ever identified, such very short-lived states would belong in the atemporal/prototemporal interface.

○ Analogous arguments hold for the interface between the prototemporal and the eotemporal worlds. Almost all of the estimated 10^{56} grams of the knowable universe is in the galaxies. The large scale structure of the universe is that of radiation, gases (particulate matter or dust), and massive matter. A whole universe made up of evenly distributed matter, of the average size of watermelons, might correspond to a world between the prototemporal and the eotemporal integrative levels; but the cosmos is not so constructed. Our formalism reflects the distinctness of dust and massive matter. Our equations handle the statistical behavior of particulate matter, as well as the macroscopic behavior of bulk matter. Statistical descriptions of space dust do not become wrong for large masses, but they do become impractical and hence irrelevant. Thus, the prototemporal/eotemporal interface is as "unpopular" as the atemporal/prototemporal one.

○ Between the eotemporal and biotemporal worlds we cross the region of biogenesis. In chapter four we considered the coming about of life and noted that whatever intermediate forms between inanimate and living matter might have existed, either evolved into more stable organisms, were devoured by these more complex creatures, or else collapsed back into the inanimate world. There are forms of life, such as the tobacco mosaic virus, which at times freeze into apparently inanimate structures, at other times behave like something alive. However, they are not examples of stable intermediate states between life and no life, but rather creatures with long evolutionary history which reproduce their kind by DNA. We did find, however, that the link that connects two consecutive life cycles, the gene, determines a rather ambiguous Umwelt between the eotemporal and the biotemporal. Perhaps our present genes are the sophisticated progeny of the structures now extinct, that once populated the eotemporal/biotemporal interface.*

○ In chapter five we found that the mind classifies external bodies, its own body and even lower mental functions as those of the non-self; in its turn, the body of man often behaves as though the demands of the mind were orders given by an alien and external potentate. Thus, there is a region of functions too mental for the body and too biological for the mind. We may think of these functions as forming a buffer zone between the mind and body and identify them collectively as unconscious psychic activities. Although unconscious processes are in an ambiguous position with respect to their stable neighbors, nevertheless they are identifiable with relative ease. We found that their Umwelt is mostly eotemporal, and we posited the disagreement between two imaginary actors of stable character as responsible for the functions of this interface: the Observer with its paleologic, and the Agent with its nootemporal world.

I propose that the metastability of functions and processes that make up the interfaces attests to the relative rapidity of the developmental phases to which they correspond.

For instance, the decay of the primordial fireball was very rapid; a few seconds, perhaps, compared to the twenty billion

---

*Recently animal cells have been successfully fused with plant cells to form interkingdom protoplast. Examples include the fusing of human cells with tobacco cells, tobacco cells with rooster cells, and human cells with carrot cells. These microscopic hybrids are not recognizable as either animal or vegetable cells, and have been playfully dubbed "plantimal" cells. Although they do not occur naturally, once created they are a potentially stable life form that would reproduce through their DNA. Hence, interkingdom protoplast is not metastable in the sense here understood, and it is certainly not between matter and life.

years of the age of the universe. This extreme rapidity is manifest today by the absence of temperatures that correspond to that of the fireball.

Going on to higher interfaces, the rapid emergence of life (in terms of the age of the world then) is the diachronic corollary of the paucity, or perhaps total absence of present forms between living and non-living matter. The relatively rapid emergence of man is the diachronic corollary of the synchronic fact that our unconscious is metastable. None of these facts challenges the validity of inorganic and organic evolution; they suggest only, as already stressed, that intermediate forms tend to evolve rapidly into more stable forms.

This sketch reveals two systematic trends among the metastable interfaces. One is that illustrative examples become more easily available as we advance to higher integrative levels. The other is that the evolutionary developmental phases to which these structures and functions correspond (the gene to biogenesis, the unconscious to the appearance of man) were brief transient periods compared to the life spans of the stable forms whence they arose. The suggestion comes to mind that in the policy of evolution the role of metastable functions is that of securing the distinctness of temporalities.

(b)   New Uses for Old Functions

Certain functions and structures of an integrative level reappear in new uses and new forms, in the features of a more advanced level. Such acts of creation abound in the regions that I described as interfaces.

○ Biogenesis is rich in examples. We have followed earlier the emergence of the cyclic order of life from inanimate clocks and found that this change brought with it the necessary "now" of life, as well as the polarization of pastness and futurity. Deterministic connectivities changed to final causation and the action-reaction principle of the eotemporal world gave rise to the mutuality of

M. C. Escher, *Day and Night* (1938). Courtesy, Escher Foundation, Haags Gemeentemuseum, The Hague.

CERTAIN PROCESSES IN NATURE cannot be classified as entirely appropriate for one or for another stable integrative level. Their positions are ambiguous with respect to two adjacent levels of complexity, such as those of matter and life, life and mind, or individual and society. Such functions and structures make up the *metastable interfaces* of nature. As a class, metastable interfaces show certain regularities regardless of the particular two levels between which they may be positioned. They comprise time's rites of passage: they give rise to new temporalities, causations, and languages (laws).

This woodcut by M.C. Escher shows in visual metaphor some of the principles that we identify in intellectual metaphor.

The figures along the vertical center of the woodcut are metastable: sometimes they seem to belong to the stable nightscape, sometimes to the stable dayscape. As we leave the lower, physical strata and concern ourselves with living matter, illustrative examples of metastability become richer, in the picture as well as in the theory.

We note in the theory that as evolutionary development crosses an interface, certain undifferentiated structures and functions on one integrative level emerge as distinct features of a new level. In the woodcut, certain undifferentiated structures of the nightscape emerge as distinct features of the dayscape. A corollary to such acts of creation is the semipermeability of the interfaces to the languages which they separate: the higher languages subsume the lower ones. The laws that determine the Umwelt of an integrative level are often unintelligible in terms of the laws of the lower integrative level. The light may shine in the darkness, but the darkness cannot comprehend it. The fine details of the dayscape are absent in the homogeneity of the nightscape.

adaptation. These new uses are quite different from, yet contiguous with, old practices. The entropy increasing and minimizing principles of the eotemporal world became the principles of decay and growth of life. The opposition to increasing disorder, manifest in physical processes as passive functions, develops into active opposition to increasing chaos, manifest in the open systems of life and self-organization.

○ Crossing from life to mind we found that the nervous system, employed by animals mainly for the control of locomotion, evolved into the human brain with functions of a much more complex nature. The auditory loop which, in animals, assists the survival of the species by such means as the blending of individual sensitivity to friend, foe, and food into collective sensitivities evolved, in man, into a means of giving a process description of the brain through language. Whereas in animals the nervous system influences the development of bodily structures through such means as sexual selection, in man it also makes possible the definition of personal identity, the symbolic transformation of experience, and the collective organization of societies.

○ The policy of finding new uses for old structures and functions is also identifiable in the physical world, although the character of its "imaginativeness" is quite different from that found in biogenesis or in the emergence of man. Consider, then, that as the universe began to expand and cool, the present structure of galaxies came to be formed. As I mentioned above, of the estimated total mass of 10^{56} grams of the universe, almost all matter is concentrated in the galaxies. The forms of energy in the primordial witches' pot ("Double, double toil and trouble; Fire burn and cauldron bubble") changed almost entirely to the astronomical universe of diminutive galaxies separated by vast and almost empty intergalactic distances ("the eternal silence of these infinite spaces terrifies me").

The eotemporal character of the astronomical universe is determined by the masses concentrated in the galaxies which take up only one-millionth of the total available volume but contain local aggregates of matter with densities of 30 orders of magnitude (and more) above the average density of the universe. This is surely a creative new use of the structures and functions of the prototemporal and atemporal worlds.

I suggested earlier that, in the policy of evolution, the role of the metastability of interfaces may be that of securing the distinctness of temporalitites. Let us now enlarge and strengthen this conjecture by adding that the role of new uses for old functions and structures appears to be that of securing continuity between adjacent integrative levels.

(c)   Natural Selection of Forms

I would assume that the forms of matter and the principles that characterize a new integrative level are selected by means of natural selection, working on the large variety of forms and principles available on the lower level. I would take this to be a special case of the more general phenomenon of morphogenesis.

○ Since the matter of the present universe is almost entirely in the galaxies, the cosmic abundance of the chemical elements is whatever we find in the galaxies. This relative abundance is remarkably uniform across the known universe, including our own galaxy and even our earth. Whereas the intergalactic spaces are occupied almost entirely by photons, the galaxies are known to comprise elements up to and above atomic number 80, and are suspected to comprise all known elements. The lightest elements (the isotopes of hydrogen, helium, and lithium) were probably made soon after the expansion of the universe began, and before the formation of the galaxies. Then, the middle weight range, up to about nitrogen and oxygen, were made by nuclear fusion-processes in stars. The origins of the heaviest elements of the periodic table are not known.

To say that there are good physical reasons for all these, is a tautology. The interesting fact which remains is that the post big-bang universe shows preferences for certain morphology and taxonomy, favors some forms of energy, and restricts others. Thus, the cosmic distribution of elements, per number of atoms, is about 85% hydrogen and helium, perhaps 5% oxygen and carbon, with the rest of all the elements taking up the remaining 10%.

O Crossing from the eotemporal to the biotemporal level, we find that life processes are peculiarly "class conscious" in the way they selected their material from the universe at large, and from the earth, in particular. Terrestrial vegetation, per number of atoms, is about 60% hydrogen, 26% oxygen, 9% carbon, with the rest of all the elements taking up the remaining five percent. This distribution does not correspond to the cosmic abundance of elements (or, which is almost the same, the earthly abundance of elements).

O Consider next the transition from animal to man. Of the almost million and a half named animal and plant species, another (estimated) million and a half unnamed species, plus innumerable other species which might have died out, it is only a very few, perhaps only a single one that came to evolve the brain whose functions determine the nootemporal Umwelt.

The initial forms of new integrative levels seem to have been "recruited" from those available on the lower level, then modified in many and substantial ways. As in the more familiar examples of evolution, the demand seems to have persisted for large stores of variations from which selections were made. The wealth of these prior integrative levels never seems to have vanished, however. Although on each level certain functions and forms are seen to predominate, at no level are these predominant features exclusive of all other features.

(d)   Change in Language

I shall argue that the interfaces are semi-permeable to the languages of the integrative levels which they separate.

Consider first a somewhat restricted use of the concept of language, that of natural language. The specialized classes of symbols, such as those of music, psychology, biology, physics, etc.

form subsets of, among others, the English language; they are not richer, but rather poorer than the domain of English language. An analogous situation exists among the hierarchy of level-specific languages.

Consider next a more general meaning of "language." This would include the classes of all signs and symbols in which the laws and regularities of nature must be expressed so as to satisfy the critical and practical intelligence of man, the formulator and tester of those laws.* It would further include all means of communication through which things and events are joined along the various integrative levels of nature. By the extended Umwelt principle one would indeed expect these languages to be Umwelt specific. By the principle of nomogenesis one would further expect that level-specific languages must incorporate all restraints unique to their specific integrative levels, while subsuming all lower order languages. Each language, then, may be seen to penetrate the languages of the integrative levels beneath it, but not the ones above it. This condition of one-way penetration is the "semipermeability" of the introductory sentence of this subsection.

There is a non-reductionist stance implicit in what has just been said. This belief cannot be justified by rigorous proof; it can only be assumed as a working hypothesis, illustrated, and in due course, made to appear increasingly plausible.

For instance, the laws of planetary motion are exemplified by the gravitational field equations of Einstein or by Kepler's laws. They are expressed in language appropriate for the eotemporal world. They could not be stated in any practical way in the language of quantum theory, which is appropriate to the particulate world, even though each particle of a moving planet may be validly described in prototemporal language. Or, let us think of the Weber-Fechner law of intensity of sensations (if one wishes to be experimental) or of the psychological regularity known as repression (if one wishes to be analytical). Neither would reveal its

---

*Sign: any object or event. Signal: any perceptible or measurable event capable of being transmitted. Symbol: any action or result of action, intended by the performer to have significance beyond itself.

significance in a language appropriate to the biotemporal Umwelt, such as in the symbols useful in studying the embryology of
vertebrates, even though both psychological laws describe the
behavior of one particular vertebrate species. One might even
argue that the laws and regularities of society must be expressed
in a language appropriate to the sociotemporal integrative level,
and are inexpressible in a language that is no richer than that
necessary for dealing with the nootemporal level.

Descriptions and explanations given in a language appropriate to an Umwelt I shall describe as *intelligible* for the purposes
of that Umwelt. Signs and symbols appropriate to a higher
Umwelt I shall describe as *unintelligible* in terms of a lower
language, while those appropriate for a lower Umwelt I shall call
obvious in terms of the more advanced languages.

For instance, the process of natural selection working on the
phenotype is intelligible in biotemporal language; the laws and
regularities that make up the gene are intelligible in prototemporal language and obvious in biotemporal language. The symbolism of dreams is intelligible in nootemporal language but unintelligible in eotemporal language.* As one explores the semiper-

---

*The fundamental discovery of scientific dream interpretation resides in the
realization that the images of the dream language are heavily informed of
biological, emotional, and intellectual needs; they do not originate from a
superhuman agency. To interpret dreams, we need not speak the language of a
divinity or spirit; it is enough if we learn to deal with birth, sex, and death in their
multitudes of roles.

The same success cannot be assigned to our interpretation of social history.
The regularities of society are likely to be those of an integrative level above that
of the individual mind and in need, therefore, of an appropriately rich language.

An isomorphic hierarchical condition is recognized in number theory. In a
study of randomness and complexity G.J. Chaitin has shown that "in a formal
system no number can be proved to be random unless the complexity of the
number is less than that of the system itself." In terms of our hierarchy of
languages this translates as follows.

If we examine a class of phenomena for the purpose of separating the
lawful from the contingent in the data (distinguishing the predictable from the
random), we may do so only if its language is simpler than the one used for the
inquiry. In that case our results, once obtained, will appear obvious. If the process
is properly that of an integrative level higher than the one in whose language the

meability of interfaces, the formal and logical implications of "intelligibility," etc., give way to descriptions of animal behavior and, eventually, to those of human attitudes. Thus, to a horse, the flowing of the water in the river might appear obvious, whereas the need to jump the fence with a man on its back unintelligible, or even mysterious. The behavior of other horses or creatures of similar complexity might appear intelligible.

In the nootemporal world of our minds it is easy to identify the emotive-intellectual equivalents of the epistemological asymmetries about the interfaces. For instance, the ways in which the self judges creatures in the biotemporal world (to wit, animals or plants) may be described in terms of indifference, curiosity, enmity, a feeling of obviousness or even sympathy, but hardly even as empathy. Empathy is reserved for fellow humans, creatures of comparable complexity; it is an exemplification of the Platonic idea that like is known by like. As our emotion about people change, so do our opinions as to which integrative level their characteristic belong. A vile character is called an animal, a good person is described as divine. The way the self evaluates the fate of mankind may best be described as one of awe, defiance, subservience, mysteriousness, or unintelligibility; the appropriate causal agents are usually elusive: historical forces, spirit, God.
    A very significant corollary of all that has been said about the semipermeability of interfaces to languages is the following. The languages of each level leave certain regions of reality and certain processes unrecognized. Using the language of a particular level, it is not possible to delineate just what these regions are. For instance, it is impossible to tell, from within the physical Umwelts (stated in the languages of physical science) that there exist the worlds of life, mind and society. It is from the unacknowledged, hence undetermined, levels that the laws and regularities of the new integrative levels arise. Because of the increasing wealth of

analysis is performed, it will not be possible to tell whether the connections among events are lawful or unpredictable. The goings-on of a higher Umwelt must remain, in this framework, unintelligible.

the languages appropriate to the hierarchy of integrative levels, the regions unrecognized on each level can be said to be systematically larger as we progress from particles of zero restmass to the mind of man. Surely one must assume that the region (left undetermined by the laws of the astronomical universe) from which the stunning variety of life arose is in some sense richer or larger than the region (left undetermined by the laws of the atemporal world) from which the chemical elements evolved.

One more manifestation of the level-specificity of languages needs to be recalled: that which leads to the uncertainty principles. This occurs whenever certain specifications applicable to lower levels are given in a language appropriate for a higher integrative level. We have encountered such principles in physics (the earliest known uncertainty principles) and in biology ("what can we mean by the birth of the whiskers apart from the birth of the cat?"). The situation is analogous when we wish to understand biological functions in language appropriate only for mental functions. Thus, when an earthworm encounters an unexpected object it tends to stop, but it does not stop "to think".

In the policy of evolution, the role of the asymmetry of languages about the interfaces may be a means (as is metastability) whereby distinctness of the integrative levels is maintained.

(e)   Conflicts and their Resolutions

It is along the interfaces, examined ex post facto, that the fiasco of prior conflict resolutions becomes evident and the leading edge of the new integrative level first becomes evident.

○ In the proto- and eotemporal integrative levels of the physical world, the principle of decay (of available energy) is opposed by a family of physical and chemical laws which assure that under conditions of dynamic equilibria the rates of decay be minimal. In the opposition of the entropy increasing and minimizing principles I perceive the workings of an inarticulate, primordial conflict. Since these specific, opposing principles vanish only if the eo- or prototemporal forms of matter, to which they apply, collapse into atemporal forms, the conflict may be said to be immanent in matter, as well as unresolvable by physical process.

In the course of the chemical evolution of the earth there came about certain substances that represent the highest degree of organization of which non-living matter is capable. But even the most efficient opposition to increasing disorganization, within the physical world, can do no more than minimize decay rates; there are no stable processes that can reverse the decay. This condition is the limitation, and therefore the fiasco of conflict resolution, available to matter.

○ Life is capable of actively opposing the decay of useful energy, expressed in the second law of thermodynamics. For small regions it can reverse this trend by creating organization and growth. Life, therefore, succeeds in resolving the conflict of inanimate matter. Continuity with non-life is retained: the opposing trends of physics evolve into the simultaneous growth and decay processes displayed by organismic functions. Since life exists only if, and as long as, the conflict between growth and decay lasts, the conflict may be said to be immanent in life, as well as unresolvable by biological process.

Throughout the career of organic evolution the incomplete adaptation of the simplest forms came to be improved through continued complexification. Implicit in this process, however, we find something which is, to use contemporary slang, counterproductive. Namely, the adaptive measures only increased the gap between what organisms expect and what they encounter, because of the very success of organic evolution. But the rate at which natural selection can produce new forms for an acceleratingly complex environment, both internal and external with respect to the organism, is limited by various parameters, such as by the range of available genetic variations. As one consequence of this situation, rates of evolutionary growth available through biological means do not seem to have been sufficient to cope with the increasing demands upon life. This condition is the limitation, and therefore the fiasco of conflict resolution by biological means.

○ The mind can narrow the gap between what is demanded of the body and what may be achieved by biological means, through its capacity to act out imaginary strategies of behavior. Humans do not need to change their biological structure, or evolve into different forms in order to adapt to changing conditions. Thus, the mind succeeds in resolving the conflict immanent in life. This remarkable feat is achieved with the assistance of a symbolic continuity, the self of man.

One model of selfhood involves the interaction of two imaginary actors: the Agent and the Observer; they correspond, respectively, to the newer and older strata of our minds. Each of these actors has its expectations, and each interprets the world as encountered, differently. The ensuing conflicts are complex, but it is clear that the human mind, or selfhood, exists only if, and as long as, this syndrome of conflicts exists. For this reason, the conflicts may be said to be immanent in the mind, as well as unresolvable by means available to the mind.

Among the functions of the mind we may identify a counterproductive trend, isomorphic to that encountered earlier by life. Namely, human creativity has produced a widening, rather than a narrowing, gap between individual expectations and possibilities. I shall argue that the vast, but limited, learning capacity of any single individual constitutes a fiasco of conflict resolution (of adaptive powers) available to the mind, whenever the individual is submerged in a collective of many minds, above a certain threshold of number and sophistication.*

*It is tempting to look to the catastrophe theory of René Thom, as the mathematical means for the handling of the metastable interfaces, recognized in the theory of time as conflict.

    Detailed examination of catastrophe theory, using the works of Thom and E. C. Zeman reveals, however, that its significance must be limited to the possible elucidation of evolutionary steps within each major integrative level of nature. The stable and continuous control surfaces demanded by Thom's theory, together with their catastrophe regions, might perhaps correspond to conditions within the proto-, eo-, bio-, noo- or sociotemporal levels, but not to conditions of nature that I

2   The Individual and Society – Part I

Using as a guide the policies identified as common to the lower interfaces, in this section I shall speculate about the likely features of the interface between the noetic Umwelt (the individual, the person, the self) and the societal Umwelt (the transpersonal, or the community of minds). In Part II under the same heading, in chapter ten, we shall deal with specific ethical and political issues in what appears as the emergent "brave new world." In this section we are concerned with the interface.

By "society" I mean any collective of humans that has existed and is likely to continue to exist for a sustained period of time, bound together by common interests that include those of survival and reproduction. By "the societal integrative level" or "the family of man," I will mean the potential, common Umwelt of most, or perhaps all people on earth.

Reasons that authorize discourse on a societal integrative level include an observation, an opinion, and an assumption.

O The observation is that the interdependence among individuals and among nations has been rapidly increasing. This growth is fostered by a universal call for a type of industrial economy which, eventually, will not be able to function satisfactorily within areas smaller than a substantial part of the surface of the earth.

There is nothing in this observation that would suggest that the future of mankind will necessarily be one of a closely knit economy, or that any such economy must necessarily involve all people on earth. It only speaks of one of the conditions toward which we seem to be headed. The other condition, as I shall argue

have described as metastable. The emergence of a new integrative level cannot be modeled by the movement of a point on a manifold, regardless of how complex that manifold may be, or how sophisticated are its catastrophe surfaces. For, the emergence of a new integrative level amounts to an increase in the codimensions and coranks of the space of the new manifold, and also to the coming about of new attractors. The new codimensions, coranks, and attractors represent dimensions and states to which nothing in the earlier world can correspond. In terms of catastrophe theory, the metastable interfaces demand transitions among sample spaces of different dimensions and sample surfaces of different character.

later, is a relapse into a fragmented world of nations, interest groups, tribes, and individuals. This uncertainty is one of the hallmarks of the nootemporal/sociotemporal interface.

○ The opinion is that if a world society were to come about, it would constitute a new integrative level of nature. We found earlier that when primordial life reached a certain threshold of complexity it was advantageous for individuals to reproduce, that is, to produce other individuals of equal or increased complexity. An isomorphic argument was used for the brain: when it reached a certain threshold of complexity it became possible for it to produce other thinking individuals of equal or superior sophistication, via the symbolic transformation of experience. In the first case, reproduction, aging, and death were born; in the second case, the mind of man was born. I believe that when the network of signs and symbols created by individual minds will itself reach a certain threshold of complexity, we may look for the emergence of a new integrative level: the societal – ready, perhaps, to produce other societies of equal or superior sophistication somewhere else in the universe.

That this opinion is not an expression of utopian yearning will become amply clear as the argument progresses. Likewise, it contains no implicit assumptions on the homogeneity of a world society. It is a conditional claim.

○ The assumption is that the interface between the noetic and the societal integrative levels will, in fact, follow the policies that lower interfaces do. This assumption does not, in itself, demand worldwide conditions; the spectrum of social sophistication around the globe can, at any moment, be rather wide. Based on this assumption, however, we shall now take up the five policies identified in the prior section.

(a)   New Uses for Old Functions

Many societal functions and structures may be recognized as extensions, across the interface, of forms well known for their roles in the mental and biological constitution of the individual. For instance, the nervous system of man, including his brain, became so well adapted to its role as the coordinator of the functions of an individual that it has not yet stopped expanding its executive domain: it coordinates groups of individuals in society. Since the brain is probably already as complex as it can be, what we have been doing during the evolution of historic man is learning how to put this complexity to better use. The societal extension of the functions of the mind is part of that learning process.

Language, (the spoken and written tongue) which we found to be the most direct process description of the brain, is the foremost symbolic link between the nootemporal and sociotemporal Umwelts. It enables the individual to formulate and establish his identity and then tie it to the symbolic continuity of the group, including that of mankind. But once language has crossed the interface, its symbols come to be defined in ways which make sense mostly in terms of the needs of the community and not in terms of the hopes and fears of the individual.

In our model of the mind we spoke of two symbolic actors: the Observer and the Agent. They represented, respectively, the evolutionarily older eotemporal, and the newer, nootemporal functions of the self. They were found to be distrustful of each other's opinions, yet it is their relationship which defined the identity of the person. In the societal integrative level we discover a perennial struggle between these two actors in the ancient roles of the Prophet and the Statesman. The Prophet is rather like the Observer: he claims to be in command of knowledge concerning both future and past, hence surveys the affairs of man in terms of unchanging criteria. The statesman is rather like the Agent. His time is asymmetrical because his future contains uncertainties with which he must deal in a crucial present. Just as with the Agent and the Observer of the individual mind, so with the Statesman and the Prophet; it is their interaction, dialogues, enmities, trusts and mistrusts that determine the creative fate of the societal body of which they are part. In the societal setting, however, the two individuals need not be, and usually are not the same person.

Agent and Observer were analytical components of a single mind. Their projection across the interface into two distinct persons constitutes new uses for old functions through division of labor. There are many other examples of this policy. Technology, for instance, is a communal enterprise that extends the bodily functions of a man in ways that no individual working alone could possible accomplish. Some such uses are not only strikingly creative but also quite unpredictable from within the lower Umwelt: there is nothing among living forms from which the scope and function of a player-piano can be foretold.

Science, as distinct from technology, is also a collective enterprise. It searches for knowledge which communal judgement holds to be true, that is, unchanging in time. Although science engages the capacities of individual minds, scientific development is unthinkable in a world of a single, independent person, no matter how ingenious he may be. Even ethics may be seen as a collective experimentation with the autogenic imagery of the individual mind, an enterprise quite meaningless except in a communal framework.

It seems, then, that political, technical, scientific and ethical practices, though originating in the individual mind, once project-ed across the noetic/societal interface, form a body of signals and symbols which are appropriate for the societal and not for the noetic Umwelt.

(b)   Natural Selection of Forms

We found that the initial forms of a new integrative level are "recruited" from a large number of candidates available on the lower level. Out of what elements should we expect the leading forms of the societal Umwelt to be selected? Out of symbolic continuities sometimes called ideas. I do not mean disembodied ghosts, but rather the actual world of symbols communicated among men. They determine an Umwelt as dis-tinct from that of the individual mind as the mind is distinct from the brain. As inanimate matter is the environment of life, the brain is the environment of the mind, so the individual minds form the environment of ideas that control the collectives of man.

E. O. Wilson remarked "that in human evolution the equivalent of an important mutation is a new idea. If it is acceptable and advantageous the idea will spread quickly. If not, it will decline in frequency ...." The mutation analogy is correct but just what is advantageous is not a simple question to answer because ideas have their own Umwelt, distinct from the biological world. Certain new ideas are acceptable, spread quickly, and in due course produce a war-torn nation of dead and cripples. We shall return to such issues in the context of ethics, in Part II.

Without detailed deliberations one cannot even guess which specific ideas might comprise the major forms of the societal world, but it is possible to make a few observations.

One concerns the identity of mankind, a symbolic continuity which cannot be defined and refined with respect to other comparable identities. Since the idea of the family of man can be defined only with respect to another cosmic image, that of the universe without man, it is likely that the selection of ideas which are to constitute the mainstay of the societal Umwelt will encounter difficulties that have no parallels along the lower interfaces.

Another observation is that for the selection process to operate on all lower Umwelts, it was necessary to have a large store of elements to work upon, even if only a few were eventually retained. It may thus be speculatively inferred that if a societal Umwelt were to emerge around a few central ideas, it will be able to retain its viability only if it remains in the company of a sufficiently large store of other alternate ideas upon which the identity of the family of man could conceivably have been built. In the words of Ivan Illich, "Any society to be stable needs certified deviance. People who look strange and behave oddly ...."

(c)   Change in Language

Modes of communication that are useful among individuals, such as the spoken, written, and acted-out languages, draw upon the biological and mental capacities of the individual. These are the languages appropriate for families, tribes, and nations. I do not believe that these languages would be appropriate across the interface, if a new integrative level were indeed to arise. I am not thinking about such trivialities as, for instance, the diplomatic language used among heads of state, but rather of the fact that individuals prefer to communicate with individuals. Franz Kafka's *K* had great difficulties in crossing the communication boundaries between his language and that spoken by the Castle.

Earlier I described the asymmetry of languages about the lower interfaces as obvious when looking down, intelligible when looking around, and unintelligible or mystical when looking upward. The replacement of individual values by collective values is often unintelligible for the individual. Again, I do not mean such practices as the willful beclouding of issues by governments, but rather the replacement of individual reality by collective reality.

For instance, it is difficult to comprehend man as a historical beast: his story is filled with more acts of senseless horror than, prorated per capita, we could admit as likely. The cruelty which members of our species are ready to mete out and the suffering they are ready to endure in the name of ideas, are nothing short of incredible. Theories that propose to account for civilizations and their discontents in terms of the drives and pains of the individual, might well reach the sources of these discontents but, because they deal with individuals alone, they cannot account for that qualitative change, something more than simple amplification of individual behavior, which is displayed by the historical motion of masses of men. The regularities of mass behavior are those of the societal and not the noetic integrative level.

Issues similar to these have been debated in philosophy under the rubrick of holism versus individualism. In sociology, they are discussed under the heading of "unintended results." Rather crassly, this is the problem of the soldier looking at his comrades: "If neither you nor I really want to be here, and if the people across the trench do not want to be here either, then why are we here?" In biology, the same phenomena is sometimes regarded as a multiplier effect: "the amplification of the effects of evolutionary change in behavior when the behavior is incorporated into the mechanism of social organization."

What we have learned so far suggests that the sources of unintelligibility, and of possible feelings of mysteriousness, are to be sought in the epistemological asymmetry of the languages that surround the noetic/societal interface. That is, explanations of the regularities indigenous to the societal Umwelt cannot be given in languages appropriate only to the noetic Umwelt. Consider the unfolding affairs of the world and witness the increasing multitudes of people, increasingly joined by proposed solutions to their common needs. The impression is unavoidable that the messages that carry their ideas have some functions of their own, and possess a degree of freedom over and above those of the individual mind.

The asymmetry of languages is encountered not only while looking up, but also when looking downward. There are many examples of such downward looks. Claims for the inferiority and insignificance of the individual as compared with the collective self, have probably been made ever since man appeared. But it is only with the explosive growth of communication, and the beckoning development of a human community, that the awesome ramifications of this lopsided relationship come into full view. Certainly, whatever is beneficial to the societal integrative level does not necessarily benefit the individual, and vice versa. This asymmetry has often been noted, bemoaned, or praised. And by necessity it has been expressed in language intelligible to the individual, for it is to him that such thoughts have been traditionally addressed. It would appear, however, that if, and as, a substantial segment of humanity enters the interface between the noetic and the societal integrative levels, traditional languages (in the general meaning here implied) will become increasingly useless for dealing with problems that center on societal identity.

(d)   Metastability

We have seen that while the processes that constitute the

interfaces are all "unpopular," they do become more noticeable as
we rise along the integrative levels. We could expect, then, that
the noetic/societal interface should be quite evident if we but
knew where to look. I suggest that, as one possibility, we look in
the direction of ethics.

Throughout the known history of man, instructions for
praiseworthy conduct seem to have existed, regardless of the size
of groups. That such instructions have always been needed
suggests that conflicts between the interest of communities and
that of individuals have accompanied the social history of man. In
the social life of animals we can also find rules for praiseworthy
conduct, very often very strong ones; but collective commands
seldom elicit what might resemble individual heroism. In the few
cases where individuals sometimes actively oppose collective
actions, such as in the case of certain social insects, no apparent
benefit, inspiration, or creative strength seems to accrue to the
individual in return for its opposition to the community. In stark
contrast, one way to interpret the history of man is in terms of the
perennial struggle between individual and communal destinies.
The reason for the profound difference is surely the absence of
animal identities (symbolic continuities in the animals' "minds")
that could find themselves opposed to the dictates of the collective
selves.

The absence of creative struggle (or tragic conflict) is not
necessarily equal to a negation of the role of the individual in
animal societies. In fact, one difficult question of sociobiology is
this: at what point does a society become so nearly perfect that it
ceases to be a society? How does one distinguish a zooid from an
organ? Some castes of certain termite species are so specialized
that they function as hardly more than organs in the body of the
colony, a superorganism; whereas siphonophores (an order of
hydrozoans that includes the Portuguese man-of-war) are both
organisms and colonies. "It is always to the advantage of the
species," according to E. O. Wilson, "to evolve new castes until
there are as many castes as contingencies, and each caste is
specialized uniquely on a single contingency." Joseph Needham
wrote in 1944 that "if any form of society is most in accord with

what we know of the biological basis of human common life, it is a democracy that produces experts."

If the biological experience is any guide, and it ought to be, the preparation for a world community might be recognized, from among other cues, by the emergence of specialization to such a degree that the individual becomes undefinable, except in terms of its role in society. It is not surprising, then, that the most acute problem of modern man is his identity – for it is of his identity that the emerging societal integrative level wishes to strip him. In traditional societies it was ordinarily much easier to be a rugged individual, living in the Norse woods, or an essentially nameless member of a human anthill (small or large), than to be both, simultaneously. It is not easy to give the King what is the King's and to the individual what is the individual's, for they are often mutually exclusive.

In modern times much thought and energy has been expended in the socially advanced portions of the world on the establishment of a working order where individual and collective interests may be harmoniously reconciled. But if the policies of lower order interfaces are useful guides, a balance between the individual and the societal selves is at best a precarious and metastable condition. As in biogenesis, or in the rapid emergence of man from his animal ancestors, the prognosis for the duration of this developmental state is one of brevity. As were the first living things which did not survive at all, and as is the buffer zone between body and mind that survives as our unconscious, a balanced condition that must characterize the nootemporal/sociotemporal interface may be only very short lived developmentally, even if it is to survive structurally.

VIII Paradoxes of Time

A PARADOX is a statement which appears as self-evidently true and without contradictions, but from which, by apprently faultless reasoning, at least two mutually exclusive conclusions may be drawn. At least since the Presocratic thinkers, many ideas connected with time have been recognized as paradoxical. Subsequent generations have slowly elaborated many of these ideas in terms of their preferred modes of explanations, yet some of the classical problems have not lost their paradoxical character. We are not interested here in historical interpretations. Instead, I shall try to state the issues briefly and clearly in ways that are acceptable in our epoch, and intend to show that the sources of some of the famous paradoxes are in the confusion that results when the hierarchical nature of time is neglected.

1 Permanence and Change

Since the Eleatic debates at the turn of the 6th century, B.C., many convincing reasons have been given for the exclusive identification of time with change (and permanence with a mirage of our minds); equally convincing reasons have also been given for the exclusive identification of time with permanence (and change with a product of our imagination). I wish to illustrate these views in contemporary form by reference to two extreme, but not unreasonable, views from epistemology.

Consider first that the universal validity of scientific understanding stems from the awesome power of mathematized knowledge. The equations of exact science provide reliable predictions about future conditions (albeit, sometimes in statistical form) when present conditions are known. The reliability of such predictions constitutes a major and necessary test of their scientific validity. Though admittedly our knowledge of nature is incomplete, there is no restriction in hard science that would, in princi-ple, delimit what is rigorously knowable. The body of mathema-tized laws stands for permanent Platonic forms that are unchang-ing through time. These conditions underlie the totality of crea-tion. What, to our mind, appears as change in time, is (in this view) but the consequence of extremely complex interactions of otherwise permanent laws; we simply do not yet know how to build up a world of apparent change from ultimate, unchanging truths. So, it seems, there ought to be no serious doubt that time is permanence.

Consider next the same scientific laws and imagine a very cold bottle, containing pure hydrogen chloride in liquid form. I can certainly make a case for the permanence of the molecular configurations of this liquid, but, as it were, only "until further notice." It is surely but a matter of time before our understanding of chemistry will have so changed that my present knowledge of permanent chemical bonds will be seen as poorly informed, good only as a crude approximation to the truth. But even now I cannot make a substance whose purity corresponds to that of the idea HCl. And if I could, I could not store it, because in ten micro-seconds, in ten picoseconds, or in ten billion years it would certainly become impure. Thus, not only am I unable to make a substance to correspond to my pure Platonic forms, but those forms themselves change. In this view, our world is one of cease-less change and what, in our thought, appears as permanence in time, is but the consequence of the intricate workings of our mind which we do not yet understand. But, there ought to be no serious doubt that time is change.

Although the reasoning just given is straight forward, and although more examples supporting either of the extreme views may easily be given, it is possible on both sides to point to weaknesses in reasoning and thereby make all arguments appear doubtful. It is tempting, therefore, to search for a theory of time in which permanence and change coexist on equal footing, but pursuing any example in search of the boundaries between time-as-change and time-as-permanence leads, sooner or later, to an unmanageable, hence suspicious, tangle. Time resists identification with permanence alone, with change alone, or with permanence and change on equal footing, a fact which has its roots in the psychological organization of the formulator of paradoxes of time.

We noted earlier that the Umwelt of the infant is one of continual change without permanent identities. As his faculties mature, the simpler perceptive modes of unproblematic change are enriched by a multiplicity of enduring identities. During this development a sense of permanence is added to an organic sense of ceaseless change.* The separation of sensory experience into components of permanence and change, as we shall see later, is a necessary corollary of quantification in terms of one and the many, of all the consequent capacities for quantifying experience, and even of the capacities for breaking down experiences into feelings and symbols. But a variety of physical, physiological, and psychological factors, heavily weighted by cultural filtering, determine how this separation is achieved.

For example, under suitable experimental conditions a changing image may be apprehended as a rotating loop of steel wire, or as a rubber loop compressed from a circle to a straight line via an ellipse, and relaxing back to its circular form. The "real thing" we would say, is either a permanent steel band moving, or a permanent rubber band changing shape. Again, another image may be seen as a steel sphere disappearing in the distance or as a balloon being deflated. The "real thing", we would say, is either a permanent sphere changing position or a permanent balloon changing size. In each of these two examples, final preference depends on information extraneous to sense impression.

In the formalized domain of thought we can think of the bifurcation of time in kinematics. Newtonian theory separated the permanent from the changing (the lawful from the boundary

---

*Marcel Proust gives a beautiful account in *Remembrance of Things Past* of the archaic experience of waking from sleep and grasping for the reality of the waking state, as identities are introduced into the Umwelt of the waking man. See footnote on p. 288.

conditions, being from becoming) in one particular way; relativity theory does it differently. Each department of systematic knowledge separates the necessary from the contingent according to its own rules of evidence.

J. J. Gibson studied the perception of moving things and changing processes and concluded that the conventional view, that the brain computes objective information about the external world gained from sense impressions, is inadequate. He maintains, instead, that the aggregate of sense organs, coordinated through the nervous system, are active contributors to what we isolate as objective invariants. But this, then, brings into the determination of "what we see as unchanging" many functions which are not normally regarded as parts of the perceptive process: language, value judgement, social conditioning.

Consider, for instance, the power of cultural filtering through language. It influences our habits of thought to a degree which is all but unimaginable by people who speak only one language. Propositions that oppose those ingrained through language are often summarily discarded as unacceptable on the basis that they are against common sense and, clearly and evidently, at variance with sense impression. We shall now consider this issue in some detail.

The modern theory of syntax is built on the reasoned belief that there exists a set of rules by which a child, learning to talk, can generate an infinite number of meaningful new utterances. This set of rules, known as the deep structure, pertains to those aspects of language which have been judged by past generations of speakers as necessarily permanent. They are continuities identified by past speakers of the language. By contrast, there must be a part of language which is totally unpredictable and unique, something that pertains to events in the speaker's life and thought that are unexpected. Only the permanent structural features combined with contingencies can make a meaningful new message. But the specific instructions on how to separate the permanent from the unpredictable differ greatly among languages.

The comparative study of languages pioneered by B. L. Whorf demonstrates that what we judge to be an event, and what to be a condition or a thing that lasts, are determined by what our language classes as a verb and as a noun, respectively. Different languages and cultures slice time into permanence and change differently. The linguistic relativity of what comprises "being" and what constitues "becoming" is a clear warning against identifying as final reality any specific separation of time into permanence and change. But the practice of division, although in multitudes of idioms, is universal among man. Its ubiquity suggests an underlying common instinctual behavior which is directed and influenced, but has not been created by social guidance. Most likely, it was inherited from simpler forms of separation into the expected (being) and the unexpected (becoming) in the life of our ancestral organisms.

For our present purposes we need not go as far as the origins of life, but only as far as the world of the infant. Connections among events in the Umwelt of the infant child are either unpredictable or, at best, probabilistic, as though everything were in the service of an overwhelming power whose purposes we might discern but could not hope to understand. We learn about such powers, as embodied by "mother," "father," "nature," and "deity." I believe that there is an instinctual drive in the adult for the reestablishment in the mind of those early Umwelts which were characterized by unpredictable events. But let us recall further that the maturing infant learns to distinguish enduring entities, and builds for himself a universe of one and the many. Throughout mature life enduring entities do change in significance and in affective charges, but contancy as such remains of utmost importance. I believe that there also exists an instinctual drive to reestablish in the mind those early Umwelts whose hallmarks were permanent certainties, first conveyed to the infant by the parent. Our instinctual drives for the identification of permanences and changes inform the whole structure of human knwoledge.

In the perceptual and cognitive development of the individual, as well as in the evolution of the physical world, the more primitive Umwelts are those of the atemporal (chaotic) and the prototemporal (probabilistic) connections. Then follows the eotemporal world whose identities endure in an Umwelt of pure succession. Because of the hierarchical organization of nature, the eotemporal conditions must subsume the proto- and atemporal conditions. But then it follows that time as change (unpredictable or probabilistic) is ontologically prior to and more primitive than time as permanence. Or, by a corollary proposition, time as change and time as permanence are not mutually exclusive conditions, nor are they on equal footing. Rather, they are hierarchically ordered, with permanence subsuming change.

Our ordinary experience of time is one of unity: time embraces everything that was, is, or will be. Problems arise when we try to identify time intellectually, either with permanence or with change; it is thus that the idea of "time" itself can be said to be paradoxical. The contradiction is removed, however, if time as change is understood as describing a more primitive Umwelt than does time as permanence, with the latter including the former. The atemporal and prototemporal Umwelts are nested in the eotemporal and higher Umwelts of our mind, and are separated instinctually by our psycho-biological faculties.

2   Motion and Rest

That there is anything paradoxical about the condition or idea of motion is not immediately evident; one must work up to it through the paradoxes of Zeno of Elea. These were formulated some twenty-five centuries ago for the purposes of discrediting the belief that multiplicity and change are fundamental attributes of the world, and promoting the idea of being and permanence as the true basis of reality. To gain his end, Zeno manipulated concepts and mental images of certain opposites, such as motion and rest, the finite versus the infinite, continuity versus atomicity, with the proverbial shrewdness of the Greek merchant.

At each instant of time, one would say with Zeno, a flying arrow must occupy a region of space equal to its length and not longer. It then follows that a flying arrow does not really move, our sense impressions notwithstanding. Our mind clearly tells us that the most important attribute of the arrow is its fixedness, its permanence, its rest. Therefore, what we call "motion" is perhaps some peculiar relationship among states of rest, recognized by our senses. The paradox is said to reside in the fact that our belief in "the flying arrow flies" is true (by our senses) yet untrue (by our logic). Since motion, at least in the Aristotelian understanding of the world is necessary for time, we may assume that for the present purposes time and motion are interchangeable ideas; hence, the paradoxical quality of motion may be extended to time itself.

The paradox of the flying arrow has been refuted, praised, ridiculed, and solved in each epoch according to the prevailing metaphysical and intellectual moods. Since about the middle of the 19th century the increasingly more sophisticated machinery of mathematical logic has been employed to refute it formally. But, as we shall learn later, number, and what one can do with number are appropriate only for issues that pertain to the lower Umwelts, whereas the paradox of the flying arrow involves the functions of our perceptive and noetic faculties. Thus mathematical solutions can give only a partial answer to this curious puzzle, even at the very best.

Let us again take a developmental approach. The concept of a motionless animal in a static environment has hardly any biological significance. Frogs will starve with a motionless fly in front of them, for the frog's eye does not tell the frog's brain that there is anything there. A jackdaw does not know the form of a stationary grasshopper at all; it is adapted to the moving form only. Sometimes, pretending to be dead, the prey succeeds in vanishing from the sensory world of the predator. Even for man, stabilized images on the retina fade away completely: the eye is primarily a motion detecting device.

Marcel Proust once described the morning discovery of himself.* As he emerged from the creature present of his dreams (from an Umwelt of ceaseless change) he slowly came to grasp the

*See p. 288.

certainty of permanence in processes and things. Through these steps permanence was added to an earlier world of pure change. Going from change to permanence, as we found in the preceding section, is a developmental step along the hierarchy of integrative levels. But if that is so, then motion itself is also prior to rest, just as change is prior to permanence. Indeed, we can combine conditions of relative motion to produce rest, but there is no way of combining relative rests to produce motion. Rest is a more advanced, less fundamental feature of our Umwelt than is motion.

The numerous discussions of this paradox that I was able to locate fall into two main categories. One type of approach insists that motion cannot be reduced to rest. The other insists on building up motion by some means, from elements of rest. The tacit assumption implicit in both is that rest is somehow more fundamental than is motion, therefore the correct path to follow (if possible) is that of reducing motion to rest. This common stance is consistent with Zeno's intentions and with the scientific taste of our epoch which, in its Platonic spirit, sees final reality in the permanence of mathematical structures. But it is an erroneous view, for the exact opposite is the case. In the economy of the physical world, among modes of perception, and among ideas, it is motion that is more fundamental and more primitive, than is rest.

Although most sighted animals would get out of the way of a flying saucer or, at least, would react to its approach, only the most advanced organisms react to resting saucers. Zeno's paradox of the flying arrow ought to be called "The Curious Case of an Arrow at Rest." Zeno's arrow always flies even if humans and some higher animals are capable of perceiving an identity called "arrow at rest." In the hierarchical structure of the world, the permanent identities of the eotemporal and higher Umwelts subsume the ceaseless motion of the atemporal and prototemporal worlds, but not vice versa. Motion and rest form an exclusive disjunction only by rank. "If rest, then also motion" but "if motion, then not necessarily rest."

3   **Beginnings and Endings of Time**

Questions of beginnings and endings of time are discussed, almost without fail, in terms of an imagined contrast between time and the timeless. The time of ordinary discourse is human time: in our terms it is nootemporality. It is generally maintained that a beginning or an ending of time must be assumed to be an event, or else it would have no character whereby it could be recognized. The paradox arises because an event, to qualify as the beginning our ending of time, would have to be "temporal" as well as "timeless," both in an unexamined sense.

Proposed solutions have been bedeviled by directing themselves to the following issue. How are we to explain the harsh contrast that, we feel, *ought to be meant* by the time/timeless interface which must surely be found at a beginning or an ending of time? I shall try to elucidate the problem by making the question more precise: What are we to mean by the beginning or ending of different temporalitites?

In an atemporal Umwelt questions of time/timelessness cannot arise; beginnings and endings can be given no meaning.

The prototemporal world is that of countable but unorderable particle-events. The lawfulness of this Umwelt comprises certain irreducible uncertainties as regards time and place; hence, no epoch or instant may be fixed with anything more precise than a distribution function. Furthermore, no meaning can be given to presentness, pastness, or futurity. One can think of emergence or submergence of prototemporal conditions as beginnings and endings of sorts, but their temporal positions are irreducibly uncertain, and beginnings and endings are indistinguishable.

In the eotemporal world the quantum mechanical uncertainties vanish, and hence events should be identifiable in time with some precision; but since the world is still that of pure succession, there is still no difference between a beginning and an ending. Thus, beginning-ending conditions must be understood with reference to other "begendings," connected by a two-way deterministic causation. Processes go neither forward nor backward, they just "go."

With the appearance of life emerges nowness and the possibility of picking a preferred direction from the pure succession of the earlier world. If all life were to vanish from earth this would be an ending of biotemporality, though not of time.

The births and deaths of individual organisms are surely beginnings and endings but, again, not of time itself. Also, we must remember that these well-defined births and deaths are evolutionary developments obtained through eons of gradual transition from the "begendings" of the physical world.

On the integrative level of the mind we have private cosmologies with clearly defined beginnings and endings, even if births are not remembered and individual deaths are not experienced. But just as my birth did not constitute the beginning of time, neither will my death mean its end.

Thus, there is nothing in nature that would correspond to the sharp contrast which, as already stressed, we think we ought to mean by a beginning or ending of time. Instead, we observe an evolutionary development that comprises a class of endings and a class of beginnings, both from a common, undifferentiated ancestry. Our very asymmetrical feelings about creation-as-beginning and apocalypse-(or the like)-as-ending are made up of a hierarchy of level-specific stages. They hang together in an integrated unity as does the body of man in the powerful Negro spiritual:

And the leg bone connected to the knee bone,
And the knee bone connected to the hip bone,
And the hip bone connected to the thigh bone,
And the thigh bone connected to the backbone,
And the backbone connected to the neck bone,
And the neck bone connected to the head bone,
Hear the Word of the Lord!

4 The Finitude and Infinity of Time

If I can construct a clock and start counting its ticks when time begins and stop counting when time comes to and end, then I shall have counted a finite number of ticks and time can be said to be finite. If I cannot construct a suitable clock, or else it is impossible to identify an instant when I am to begin the counting and another instant when I am to stop it, then time may be said to be infinite. There have been many and differing opinions as to whether time is finite or infinite. We shall examine this issue from the point of view of hierarchical time.

In the atemporal world there are no clocks conceivable, hence neither finity nor infinity of time makes sense.

A prototemporal clock (perhaps radioactive emission from a weak source) is intrinsically inaccurate. There is no assurance that it will tick within any predetermined amount of time; it is analogous to a rubber yardstick that may be stretched to any desirable length. It follows that a determination whether or not a prototemporal Umwelt is finite or infinite (in time) does not seem to be possible – quite apart from the fact that prototemporal distances and times are only weakly differentiated.

Eotemporal clocks ought to be reliable, that is, their readings may be mutually compared according to set rules. But time is still characterized by pure succession: there is no present, no futurity, and no pastness. Hence, there is no way of telling whether a series of counts is to be added to or subtracted from other counts. In a curious way, therefore, finity and infinity seem to mean the same thing.

Biotemporal clocks may also be reliable, and their ticks are certainly countable. Since life does define a nowness, as well as futurity and pastness, clock readings may be systematically accumulated so that we may determine whether time is finite or infinite. But alas, we have other problems: when do we truly begin and when do we stop the measurements? The issue of finity versus infinity joins here the issue of beginnings and endings discussed in the prior section.

The nootemporal world is that of private cosmologies, wherein the life of an individual is surely finite. We can imagine and postulate the existence of time before birth and after death, though it is impossible to demonstrate to an individual that such a continuity is indeed the case. We encounter here a dichotomy, a contrast between the self and everything else. The self is finite and the world apparently infinite in time. But as we descend along the temporal integrative levels the mutually exclusive conditions of finite and infinite time become indistinguishable, and in the atemporal the issue disappears.

5   Atomicity and Infinite Divisibility of Time

By infinite divisibility of time we mean that, no matter how short a period we may select, it is always possible to divide it into even shorter periods. The idea of infinitely divisible (or continuous) time is isomorphic with the mathematician's continuum of real numbers, and hence, with the geometer's continuum of points on a line. Numbers in a dense continuum are, of course, not numerals (different symbols) but ideas of "how many-ness." Likewise, points on a line are ideas of "pointness." Thus, whether in mathematics or geometry, the elements of a dense continuum are featureless in that they are indistinguishable; they form an aggregate smoother than the best vanilla ice cream. It is the complete absence of structure and distinguishability which permits the set of real numbers to be dense.

In chapter one we defined an atemporal Umwelt as one in which nothing can correspond to any of the hallmarks of time. In chapter two, beginning with this definition, we identified many atemporal conditions and classed them all as "chronons." Then, an "event" was defined as any thing or conditions that remains identical with itself through a period of time. All chronons are events and all events are chronons (and the two words are retained only to secure continuity in their historical uses). The idea of an (atemporal) event is therefore isomorphic with the idea of a dense continuum: both are featureless and hence offer no resistance to division without limit. Also, any chronon contains as many points as any other chronon, that is, a high infinity of them.

We identified events as the atoms of time, because all higher temporalities are constructed from chronons. Their relationships create the principles of temporal hierarchy: countability, arrangeability, nowness, preferred direction, and so forth. The conclusion is then reached that atemporality is continuous, or infinitely divisible, but also that all higher temporalities are atomic. As in the other paradoxes which we have discussed so far, so in this one, the apparently opposing attributes (continuity and atomicity) are found to be hierarchically related, with the atomicity of higher temporalities subsuming the infinite divisibility of its atemporal atoms.

Mathematical physics often and successfully employs the idea of mathematically continuous time for problems which, in our framework of thought, pertain to temporalities above the atemporal. All such applications, however, exemplify only the well known practice of theoretical physics, wherein logically fictitious devices are employed for practical efficacy and mathematical convenience.*

The idea of atomism, as I stressed earlier, closely relates to the logic of explanation: what do we explain by what. If atoms of some kind are to be used for the elucidation of a structure or a process, then the atoms themselves, from the point of view of explanation, cannot comprise other structures or processes, for then they cease to be the fundamental elements of explanation. Thus chronons can serve as atoms of time, but, as we also learned in chapter two, they cannot be placed end-to-end so that they add up to continuous time. They have no ends, and only they themselves are continuous, not the higher order phenomena built upon them.

6   The Unexpected Egg and Its Chickens

The problem of the "unexpected egg" is a powerful paradox of time, though generally it is not so recognized. By varying the

---

*For instance, a differential charge dq is a discrete unity. When this symbol is used in differential equations, it is assumed to be integrable into Q, the total electric charge, as though Q were infinitely divisible. But Q is not infinitely divisible. We employ such a convenient sleight of hand on the sole basis that dq is ordinarily immensely smaller than Q – which is true.

emphasis on the issues involved, it may be shown to incorporate the paradoxes discussed so far – and then some.

There are ten closed boxes labeled from 1 to 10 and we are told to open them, one at a time, in the order in which they are numbered. We are also informed by a reliable person that one of the boxes contains an unexpected egg. By "reliable" is meant that past opinions and predictions of this person, after they have been checked, have always turned out to have been correct. By "unexpected" or "surprising" is meant that we cannot deduce by argument alone in which box the egg is, until the ,box has actually been opened.

First, before actually looking, we shall think about the experiment (the opening of the boxes) and argue as follows. The egg cannot be in the 10th box because, were I to find the first 9 boxes empty, I then would know that the egg is in the 10th box, hence it would cease to be unexpected. It cannot be in the 9th box either because, were I to find the first 8 boxes empty, and knowing that it cannot be in the 10th, I would know that it is in the 9th box, hence it would again be an expected egg. By iteration we work our way down to the first box and conclude that the boxes are empty. Then, we begin the experiment and find the egg in any one of the boxes.

The paradox resides in this: the argument, as well as the experiment, is straight forward and unassailable, yet they lead to opposing conclusions.

Another form of the story is that of a prisoner condemned by a judge to be executed, unexpectedly, before ten days are over. The backdrop is the same. The prisoner's lawyer declares with a grin that his client cannot be shot on the 10th day unexpectedly, etc. But he is shot just the same on one of the days.

For consistency of phrasing I shall call the individual boxes or days *Intervals* (of time); the unexpected truth of the egg or of the execution, the *Event;* the person employing the recursive logic, the *Logician* or the *Lawyer;* the one seeking the truth the *Experimenter,* and the person making the original pronouncement, the *Lawgiver.*

The lawyer-Logician operates in an eotemporal Umwelt of pure succession where intervals are countable and orderable by number, but there is no preferred direction of time. Consequently, he regards reasoning from ten to one on equal footing with reasoning from one to ten. The Experimenter operates in a biotemporal or nootemporal Umwelt and must go from one to ten. The Lawyer is quite correct for his Umwelt, because there is nothing in his abstract world of real numbers that would correspond to a preferred direction of time. But the prisoner is alive, and life favors only one of the two opposing directions of time. The paradox derives from the careless mixing of temporalitites. Unlike in the deterministic world of eotemporality, in the experiential world of the Experimenter the unexpected can happen, and even the lawful but yet unexpected is a well-known phenomenon.

Let us now examine some of the chickens that may be hatched from the unexpected egg.

○ In Christian tradition the end of the world will be reached in a finite number of days. That end will be preceded by the Day of Last Judgement. "But of that day or hour no man knoweth, neither the angels in heaven, nor the son, but the Father." The God of Christianity, the superior Lawgiver, is outside time and hence he knows the answer. Lawyers, logicians, and theologians may argue this Judgement Day Paradox: "the Judgement day cannot happen on the day when the world ends for then it would not be unexpected." But man lives his fretful life in a nootemporal world. His future contains irreducible uncertainties, symbolized by the End of the World in Christian tradition, and also lawful yet unexpected events, as symbolized by the Judgement Day, and exemplified by radioactive radiation.

○ The first and last Intervals are quite different from the others, for they have only one neighbor each. Our ten boxes generate troubles, as do cosmologies when we consider the beginning and ending, and the finitude-infinity of time.

○ It is possible to have only one single box which is said to contain an unexpected egg. "Unexpected" here means a suspicion: although the Lawgiver has always been reliable, the future might not copy the past. As we seek to determine whether there is an egg there at all, we may open the box very slowly and, we are back at the paradox of intervals. Are we going to be able to see the egg when the lid is opened 1°? Or when it opens 2°? Can it be a 90° egg? or an 89° egg? As we try to divide the opening into smaller and smaller intervals we are back at the issue of infinite divisibility of time.

○ If we permit the instantaneous opening of the box, the character of the problem again changes. Either there is an egg (or an Aristotelian seabattle) in the box or there is none, but the egg cannot be either expected or unexpected. It either exists or it exists not. The box becomes an atemporal universe which cannot accommodate contingencies because, in the words of T.S. Eliot,

> only in time can the moment in the rosegarden
> The moment in the arbour where the rain beat,
> The moment in the draughty church at smokefall
> Be remembered ....

○ The Lawyer's reasoning could be applied to Zeno's arrow and the claim made that it cannot hit the target at any unexpected instant before sunrise tomorrow at 5:47 A.M. Because, if it did not hit the target by 5:46 A.M. it could not do so at 5:47 A.M. and so forth. And, since the arrow is now at rest, it cannot fly between now and sunrise tomorrow, or between now and Dooms-day, unless the world is totally deterministic.

○ The Lawmaker's pronouncement may be restated thus: "You shall hear an unexpected tick from this box before its machinery runs out tonight." In this case we must be dealing with a rather slowly ticking clock, for all clocks are devices that deliver predicted ticks at unpredictable instants. If the instants could otherwise be predictable, we would not need clocks.

○ The Tristram Shandy paradox, so named by Bertrand Russell, is yet another chicken of the unexpected egg.

In Laurence Sterne's novel *Tristram Shandy* the title charac-ter, the autobiographer Tristram, complains that it took him a whole year, and three and a half volumes of writing to reach "no farther than to my first day's life ... so that instead of advancing, as a common writer ... I am just thrown so many volumes back ... at this rate I should just live 364 times faster than I should write ..."

The mathematics of Shandy, Esq. is curious, but his concern is clear: he can never finish his autobiography. Not so, said Bertrand Russell. Since for every day lived there corresponds some day, or days, during which the events will be written up, all we need to give Shandy is an infinite number of days and he shall complete his task.*

Using the concept of transfinite numbers, the infinity of Shandy's days on earth may be identified with $\aleph_0$, the cardinal number of the set of all natural numbers. It is the property of $\aleph_0$, that any of its proper subsets equals the cardinality of the set. Hence, all that is necessary is to assure us that the number of intervals marked are in one-to-one correspondence with the number of days recorded, and Russell's claim has been proven – almost.

It is characteristic only of finite collections that the whole is never equal to a part, not for infinite collections where they both may, though need not, have the same power. Collecting every Tuesday from Shandy's infinite life gives us a transfinite number of days which has the same power as that of all weeks. But, and this is an importanz "but," there is no genidentity between days of experience and pieces of paper. There is no valid way in which the creative events of Shandy's activities (such as recording the words of his free choice) can be put in one-to-one correspondence with sets of numerals. The quantity $\aleph_0$ is a cardinal number; hence its Umwelt is prototemporal. Tristram Shandy's writing determines a nootemporal Umwelt. The Russell argument contains a fatal category mistake because it mixes temporal Umwelts indiscriminately.

---

*Second order effects are neglected. For instance, the world itself might not last long enough. Or, he will arrive at a day when he shall have to begin copying the events of the day when he began writing his diary, and this series does not converge, etc.

Those familiar with set theory will realize that what we have been doing is to distinguish meticulously between properties of sets on the one hand, and on the other hand, relations on the sets. But such insistence alone would have remained a mathematical exercise, were it not for the fact that members of sets belonging to one Umwelt can be distinguished from the sets themselves which, qua sets, determine higher Umwelts. The Umwelts, in the examples, correspond to integrative levels of nature, each with its distinct temporality.

There are no corresponding experiential paradoxes; they appear only when, in abstract language-thinking, the hierarchical character of experiential unity is neglected.

IX   Science as Truth

THE UMWELT OF AN ANIMAL comprises outward projections of organized, elementary sensations. The aggregates of signs and signals within the organism which determine this Umwelt form an internal chart of reality. In the case of man, the most significant such chart is what I have called the internal map of the mind. Regardless of the sophistication of this internal map, however, its contents may be conveniently thought of as a store of knowledge. Knowledge, so understood, is identifiable through behavior and, in the case of man, through language.

One of the functions of the human brain is the continuous enrichment of its internal map through autogenic imagery. But if a vast collection of self-generated images is to be useful to the happy owner of the mind, he will have to function with the assistance of selection rules: it is not possible to follow up all ideas that one may think of. The true, the good, and the beautiful are traditionally such selection rules. With the good and the beautiful we shall deal in the next two chapters; here we are concerned with the true.

It is surely through permanence that truth joins the issue of knowledge: only stable beliefs will be consistently useful in the "preservation of favoured races in the struggle for life." Accordingly, I shall define truth as that class of human knowledge which individual and communal judgments regard as permanent. In our epoch the foremost arbiter of truth is scientific knowledge; hence there now prevails a common cause between the idea of truth and that of science. We shall explore the origins, validity, and limitations of this common cause.

1   **The Morphogenesis of Human Knowledge**

We shall try to identify some of the ways in which knowledge is transformed from its earlier, biological to its noetic forms.
I will argue that the development of learning has its roots in the
evolution of certain functions which cross the biotemporal/noo-
temporal interface. During this process knowledge opens up, it
becomes enlarged and enriched, as do some other functions,
discussed earlier. An example will illustrate the point I wish to
make.

In chapter seven we found that one of the policies of
evolution has been to project certain practices across the metastable interfaces which separate integrative levels and, with stunning ingenuity, put old forms to new use. One such projection was
identified by Erik Erikson in his work with spontaneous play
patterns of children between the ages of 8 and 12 years.

He gave a variety of figures and building blocks to boys and
girls and asked them to construct "exciting scenes." Though his
tests were originally intended for a different purpose, he came to
realize that, quite consistently, boys in these tests erected and
constructed. With hardly any exceptions, their scenes were houses
or facades with protrusions and high towers, infused with motion,
accidents, collapsing structures and ruins. Girls' scenes of houses
were almost entirely of an interior nature: they included, enclosed, and held safely. Erikson recognized in the morphology of
the two play patterns the biological genital differentiation. In the
male, wrote Erikson, "the external organ is erectable and intru-
sive ... serving the channelization of mobil sperm cells; in the
female [the organ] is internal with vestibular access, leading to a
statically expectant ova." That there are morphological differences between the sexual organs of males and females is not a
very new observation. What is stunning, however, is that in a type
of morphogenesis from biological functions to behavior, the
spontaneous preferences go far beyond unconscious representation of sex organs by non-verbal communication: they project

themselves into life styles and, one would expect, into world views.*

We found earlier that the psychological development of man is a phased sequence whose major steps correlate broadly with the hierarchy of temporal Umwelts. These developmental levels, often connected by some such mechanisms as the one I have just described, are all represented in the psychological endowment of a mature individual. Because of the hierarchical organization of nature, the language of each of the developmental levels subsumes the languages beneath it and, as a corollary condition, each level is restrained by the lower level controls. For instance, from the biosemiotic approach of F.S. Rothschild, we learn that

> the noetic system adjusts in the oral phase to cellular system intentionality, at the anal stage to the gastrular system, to come forth in the latency period with its own style of intentionality; in the following periods, and in retrograde order, this style then rules successively over the neural, then the gastrular, and finally over the cellular system.

Yet the person comprises these levels in a hierarchical unity. While Odysseus the archer struggled with his plans and memories, his mind functioned in the nootemporal Umwelt. Simultaneously, his body aged as his metabolism followed its ancient, circadian rhythm. His gametes were deathless in the sense that they did not age; their very stable structure had been inherited from Laërtes and was passed on to Telemachus. When Odysseus jumped off his long boat he fell according to the laws of

---

*If these conjectures appear too fanciful, one should remember that the sexual act is the most important biological function of the individual, beyond staying alive. Males of most advanced species participate in this ubiquitous ritual by injection, females by reception. It is not strange, therefore, to find the morphologies of sexual organs transposed into the forms and modes of knowledge, considering that the lineage of man is some 200,000,000 years old. The number of times that sexual organs were put to use during this period is very large.

the eotemporal world. As he fell, he beheld the image of the Island of Calypso the Concealer. That image was carried as a modulation on electromagnetic waves which determine an atemporal world, and absorbed on the retinas of Odysseus' eyes according to the prototemporal processes of quantum physics. And Odysseus knew nothing about the hierarchy of Umwelts; it is we who consider these issues.

   With this hierarchical unity in mind, we wish to find out more about how biological knowledge transforms into noetic knowledge. In this inquiry we intend to take advantage of information hidden in human language, because, as we found earlier, linguistic skills cross the biotemporal – nootemporal interface.

   A practical entry into this domain of linguistically based epistemology is with the comparative examination of the psychodynamics of many tongues, a technique first worked out in critical detail by Thass-Tienemann. The principles of this approach are firmly bound to the linguistic analysis of words in all their shades of meaning and to the study of changes they have undergone through their known history.* A comparative study of several languages suggests that throughout the recorded history of their development, there has been a trend toward altering meanings that are emotionally charged, especially if they are anxiety provoking. In these languages words that relate to birth, sex, or death tend to assume new meanings that relate to abstract knowledge.** In some cases the original meaning becomes repressed. We should note here that questions about birth and

*This very recent psycholinguistic approach reminds one of John Stuart Mill's work on language, first published in 1843. If one attempts to rectify the use of a vague term by giving it a fixed connotation, one endangers one of language's "inherent and most valuable properties, that of being the conservator of ancient experience; the keeper-alive of those thoughts and observations of former ages which may be alien to the tendencies of the passing time .... Language is the depository of the accumulated body of experience to which all former ages have contributed their part and which is the inheritance of all yet to come."

*The languages involved in the comparative study are English, French, German, Gothic, Greek, Hebrew, Hungarian, Italian, Latin, Lithuanian, Old English, Old High German, Old Norse, Sanskrit, Slavic, and Spanish.

death are direct inquiries into being and not being, and hence into time and no time. The same holds for issues on sexual union, for reproduction is the foremost biological means for overcoming death.

The historical path of changing meanings may be traced backwards from present, abstract meanings to early, specific ones; that is, from the present noetic to the earlier biological significations. Through this linguistic archeology four major ways may be identified in which language reveals our capacities for acquiring knowledge. More accurately, these are the ways in which we, the makers and users of language, give an account of our unconscious awareness of the biological roots of human knowledge.*

---

*Studies of psycholinguistics hint at the active role that the perceiver plays in forming percepts. The argument is based on the observation that words, just as individual thoughts, are shaped or totally forgotten, yet these meanings remain unconscious, latent powers beneath dictionary definitions.

Thus, both *perceive* and *conceive* denote mental activities related to *grasping* (understanding) as well as to creation. The Latin *percipere* contains the idea of "taking hold" from the Latin *capere*. The German *begreifen* (grasping an idea) originally meant "grasping with the hands." For fifteen million years before the creation of a new mental concept came to be described as "grasping an idea," women were grasped so as to make them conceive new life. When our distant ancestors sought a linguistic symbol that would represent those feelings which signal the birth of a yet nameless idea, they must have been moved by a sense of kinship between the two types of conceptions: the biological and the mental. The ancient meaning survives in our present use.

The word "perception" is ambiguous. It means (1) sense oriented awareness of change and permanence in the self or in the environment, as well as (2) the internal events responsible for this awareness. This ambiguity is probably not accidental. The word reflects the ancient confusion about the reality of an external world. Bluntly put, is it "me" that I see projected out there, or is the "out there" something quite independent of "me?" Is time in me, projected on the external world, or is it out there perceived by me? The ambiguity implies an awareness that perception includes a creative stress, not unlike that of carnal knowledge.

The tacit and erroneous assumption that the organism is a passive observer of the external world has been called the dogma of immaculate perception. The position taken by the theory of time as conflict is a diametrically opposite one: it is that of passionate perception. Such a stance is implicit in the extended Umwelt principle, put forth earlier.

According to the witness of our "unconscious language," external reality may be assimilated through the mouth, explored by the genitals, investigated through seeing, or made one's own by listening. It is convenient, therefore, to distinguish oral, genital, ocular, and aural knowledge.

○ The suckling infant tastes and smells the mother and feels the smoothness of her skin; for him, sense perceptions of taste, smell, touch are undifferentiated and charged with emotions. In the course of cultural development, the meanings of these words have drifted toward secondary meanings, less highly charged. Thus, the meanings of taste, smell, or touch shifted from the sensual to the cultural. For instance, one can trace the development of the noun *taste* from its early meaning of oral experience to its cultural meaning of taste in aesthetic judgment. The English verb *to taste* corresponds to the German verb *tasten* ("to touch"). Both come from the Latin taxō, -āre, "to touch." The point is that later judgments on what exists, and what does not exist, are directed by infantile ways of knowledge, both ontogenetically and phylogenetically. Those fields of knowledge whose symbolism is oral ("sweet music," "a touching story," or a "hot performance") are likely to be without formal epistemologies.

○ The meaning of words that deal with genital exploration tend to shift from the sexual to the mental. "Grasping" and "conceiving" (as I mentioned above) for the production of new people is surely an older experience than the realization that one must grasp certain facts before one can conceive of new ideas. While taste, touch, and smell are innerdirected, genital knowledge is outward directed; its paradigm is the "carnal knowledge" of woman; its thrust is the intimate exploration of another person. The impulse that appeared on the level of oral knowledge as introjection of all things external, becomes refined and differentiated on the genital level. In its masculine form (whether in man or woman) it turns into aggressive penetration of the unknown: virgin woman, virgin forest, the inner secrets of the atom, or the hidden laws of the cosmos. In its feminine form (whether in woman or man) it grows to a new capacity for internalization of distinguishable identities: friends, Romans, countrymen.

The genital act has been perceived primarily as a cognitive act throughout Western history: "Adam knew his wife; and she conceived." This unconscious understanding can be traced along many paths to our own days. Sexual experience in humans displays a continuum between work and play, and both work and play may assume sexual dimensions. When the meaning of work (in the present context, the acquisition of knowledge) becomes particularly intense it tends to assume sexual significance and it has often been so represented in symbolic form. Creative work is indeed a continuous search for something hidden in the object of knowledge, a quest for a reality which is more than what is at hand, something which may enter conscious experience at any moment or else be revealed by being entered into.

In the words of an unknown chanticler of the fourteenth century

> And all was for an appil
> An appil that he tok,
> As clerkes finden
> Written in their book.
> Deo gracias! Deo gracias!

Shakespeare knew the story more explicitly

> But love, first learned in a lady's eyes,
> Lives not alone immured in the brain,
> But with a motion of all elements ...
> ... gives to every power a double power
> Above their functions and their offices.

As far as I can tell, all civilizations have employed complex social and psychological forces to control sexual behavior. Under their constraints, the meaning of words and concepts that relate to genital knowledge tended to drift from the biological toward the mental. Earlier meanings, as I mentioned, are often retained. Mature judgments on what exists and exists not, in those domains of knowledge which may be explored by means whose paradigm is genital, are therefore biased by the same forces that control genital behavior. Even in our day lovers, as well as judges and scientists, seek to discover the naked truth.

○ The meaning of words that deal with the visual exploration of the world tends to shift from biological meaning toward abstractions that we often associate with the Platonic world of ideas. Being "enlightened" still carries an archaic color that relates it to fire, but primarily it signifies the actions and views of a well-informed person of good will. "Light" is in many ways synonymous to "self" or "mind." When the light is extinguished, a person ceases to be a person. Light is perceived in all great cosmologies as a symbol and source of reason, while ignorance is perceived as darkness. The two are mutually exclusive as are day and night for "the light shineth in the darkness; and the darkness comprehended it not." The archaic meaning for foresight is spatial ("my eyes were weary with the foresight of so great a distance") but since at least the 16th century its meaning has come to be divorced from sight and its significance is now abstract and temporal.

The psychodynamics of this shift is not as clear as in the cases of oral and genital knowledge. It is suspected, however, that the psychological power that motivates the shift comes from prohibitions on looking, especially as upon a parent as a sexual partner.* The Old Testament controls incest thus: "Cursed be he that setteth light by [that is, observes] his father or his mother." Whatever the causes of the change in meaning, from the biological to the world of ideas, the implication is that judgments upon things and issues whose paradigm is ocular knowledge, will be heavily biased by the judgments of the superego as to what ought and what ought not to be seen.

○ Compared with ocular, genital or oral knowledge, aural knowledge has been stable. We learned earlier of the unique position of hearing in the definition of the self: though I cannot

---

*Cf. Section four of this chapter where we learn about cultures that emphasize hearing over seeing.

see, smell, taste, or touch myself in toto, I can often hear myself as all others hear me. The biological significance of hearing the sounds of an external world is still with us, though it has become enlarged to mean listening to the inner voices of the self. These tend to be emphatically intimate. We hear the call of duty: our better self commands; we may also listen to the words of the devil within and succumb to temptation. A young man may still feel he has been called to become a physician, as by a voice, rather than signaled, as by a lamp. Listening with the third ear amounts to hearing one's own self better. In Latin "to hear clearly" is *exaudire* (hearing from the outside) whereas "to learn" is *inaudire* (to hear in secret or from within).

It is interesting to take an evolutionary view of the development of the four major paradigms of knowledge that we have identified.

(1) Oral knowledge is the distant progeny of primitive sensitivity to chemicals and heat. Fields of knowledge with oral symbolism are likely to be organic rather than analytic.

(2) Heterosexuality, a technique that emerged together with death by aging, gave rise to a type of appreciation unavailable to organisms that do not reproduce heterosexually. Its characteristic forms are those of the aggressive penetration of the unknown and the internalization of distinguishable identities.

(3) What began as a means to distinguish at a distance friend, foe, and food by visual means gave birth to ocular knowledge that seeks the measured order of the world.

(4) What began as a means of distinguishing friend, foe, and food by hearing, came to be an essential means in the process of self-definition. The characteristic of aural knowledge is that of privileged disclosure that pertains to the "inner world."

Whereas oral and genital knowledge may be called Dionysian because of their orgiastic character, ocular knowledge is Apollonian for it favors measure, order, and balance. Clear thinking in the light of pure understanding is the merit of the unemotional world of geometry and the non-contradictory world of mathematics. Compassion for those who labor and are burdened is the merit of humanistic knowledge.

These four evolutionary trajectories are representative of many; they are not intended to be exclusive. But they are important because they demonstrate some of the ways in which the biological origins of the intellect have left their hallmarks on human knowledge. What we judge as existent or non-existent in our epistemologies, the things and events that populate our noetic worlds, are not as detached from our evolutionary ancestry as they are usually assumed to be. Rather, our scientific and philosophical utterances derive from biological capacities and are heavily modulated by psychological forces, value judgments, and social filtering.

2 Personality, Knowledge, and Time

The impressive historical continuity of personality theories attests to a desire for the unification of our knowledge of human behavior under a single rubric. But the scores of definitions of what is meant by personality suggest that while the goal might well be intuitively clear, the task is difficult. Since time is surely central in the affairs of man, one would think that a person's attitude to time might be a suitable index of his personality. There exists a large literature on personality and time. But a substantial portion of the writings miss a common focus; they are incoherent in approach, interest, and even in vocabulary, providing the reader only with tantalizing, kaleidoscopic information. Such a disarray often marks new fields of research that do not yet offer generally accepted unifying perspectives.

I believe that the theory of time as conflict offers such a perspective. It suggests an index of personality based on attitude to time, and provides a scale upon which the indices may be entered.

Consider first the Augustinian dilemma: if no one asks me what time is, I know what it is; if I wish to explain it, I know not. Let us call this the indeterminacy principle of St. Augustine.

According to this principle, time felt and time understood are, at any one instant, complementary and coexistent traits of the personality. The scale stretches between the imaginary conditions of time purely felt and time purely understood. At either end one mode of appreciating time would exclude the other. Thus, as I approach the intellectual extreme (a purely ocular form of knowledge) I shall end with marvelous theories from which all important aspects of my experience of time will have been banished. If I go the other way and propose to attend exclusively to my feelings about the passing of time (perhaps a purely oral form of knowledge), I am likely to end up hearing unrepeatable voices and experiencing undescribable ecstasies. The experience often called "human time" or, in our terms, nootemporality, obtains in the middle regions (perhaps in the form of genital knowledge) where the unresolvable conflicts of the mind between the phylogenetically older oral, and the newer ocular types of appreciation, the conflict between passion and knowledge, remains unresolved. The experience at either extreme end of the scale tends to appear timeless.

Personality may therefore be thought of as a gamble of balance: a peculiar manner in which an individual prefers to play his archaic logic against his discursive logic. But since only one action is possible at a time, there must prevail at each instant a preference for a peculiar mixture of oral-genital-ocular types of knowledge. As the personality of the child becomes defined and refined, one weighted mixture of attitudes is likely to be established. Though this attitude surely varies from instant to instant and from mood to mood, its center of gravity would remain, I believe, a reliable index of personality: ordinarily it would be subject only to slow change.

The personality index along the Augustinian scale involves the criteria which a person will employ for the admission of certain evidence or explanation as truth: the index shows how one prefers to divide the unity of his experience of temporal tension into components of knowledge felt and knowledge understood. As I have just stressed, consistency in people's actions suggests the stability of the passion-knowledge quotient. Some people are likely to stress the absolute superiority of intuitive understanding over the analytical, some vice versa. But it is a universal experience that making decisions and remaining consistent with one's personality involve a series of delicate balancing acts; therefore few individuals would insist on being placed at a fixed position on the Augustinian scale, to the exclusion of all others.

Fields of knowledge, however, as far as codified attitudes for evidence is concerned, need not be concerned with the weakness of the flesh. They are but aggregates of signs, signals, and symbols recorded or recordable upon members of the eotemporal Umwelt, such as books, stones, magnetic tapes, or bronze. Established fields of knowing usually insist on certain very specific ways in which the necessary must be distinguished from the contingent. They have different and well defined rules of evidence, and these rules are not negotiable without major revolutions in the preferred structures of knowledge. Departments of scientific knowledge, at the Apollonian end of the spectrum, represent idealized, unadulterated, personality preferences.

3   Number

The authority of science that makes it the foremost arbiter of truth in our epoch derives from the power of mathematized knowledge. In its turn, that power must be attributed, in Eugene Wigner's words, to the "unreasonable effectiveness of number" in natural science. In this section I shall try to identify the sources of this unreasonable effectiveness.

As a child grows, his world becomes populated with enduring objects other than, and independent of, himself. The growing child (or the evolving Homo) leaves behind his infantile monism and learns to live with the idea of one and the many. The enduring symbol which is the self, that unbroken continuity between birth and death, remains the referent for all continuities and assumes, among the "many," the privileged position of the "one." I believe that this unity of the self is the archetype of one-ness. The capacity to know the world in terms of distinguishable continuities through time made counting possible, and with counting, the quantifying of experience.*

For Pythagoras and up through the 16th century "one" was "the root of every number but itself not a number." "Two" was the least number. Odd numbers were said to have had masculine virtues "proper to the celestial Gods" whereas even numbers were indigent, imperfect, female.** Shakespeare played on the curious position of One when he made old Capulet urge Paris to compare young women at the feast:

> Hear all, all see,
> And like her most whose merit most shall be,
> Which (one more view of many, mine being one),
> May stand in number, though in reck'ning none.

Developmental psychology shows that first we learn to compare, then to evaluate. The child has to learn first how to group objects according to identities: all apples as distinguished from all children. The idea of a set is then acquired from the notion of simple ordering. Then he learns that the spatial arrangement of a set of objects does not alter the numerical value of a set; the child thus learns to deal with numbers as classes as he discovers the principle of conservation of number. In producing a class, the child masters cardinality. Within each class all members are taken as identical and indistinguishable; hence the operational Umwelt of cardinality is the prototemporal.

---

*It also became possible to abstract invariances from feelings and other sensations whose external sources could not be identified. With a private world of meaningful experience one might expect to see metaphor making emerge, and might speculate about the opposition between the deathless world of symbols and that of the inevitably aging bodies.

** The painfulness of change from the unity of the mother into the twoness of mother and child is recorded in some languages. In English *alone* comes from the earlier *all-one*, used to describe the newborn as whole and one. The experience of *loneliness* is a corollary of being separate.

Ordinality is discovered when the child is able to assimilate relationships such as "more than" or "less than." What he learns is pure asymmetry, for natural numbers arranged in a decreasing sequence are also thereby arranged in an increasing sequence and there is nothing intrinsic in the ordering that would make one direction preferable to the other. The operational Umwelt of ordinality is the eotemporal: members of this Umwelt are not only countable but also orderable by number.

The development of the child's mathematical ability recapitulates the developmental stages of mathematical knowledge. Recognition of identities are evident in paleolithic remains, followed by the capacity for simple ordering, which is a prerequisite for any written language. Out of the chaotic welter of experience there seems to have arisen the mastery of sets (sheep-to-twigs correspondences) and eventually, very slowly, the handling of number. Negative numbers first turn up in 13th century China and in 16th century Europe. Zero as a place value notation came into use during the Seleucid Dynasty of kings in nearer Asia, three or two centuries before Christ, but the use of zero, as a reference point about which two semi-infinite rays extend, was not possible until the invention of negative numbers. The sequence of mathematical development thus progresses from a formalism appropriate to prototemporality (cardinality), to one appropriate to eotemporality (ordinality). Only with positive and negative numbers and zero can we represent mathematically a "now" – of sorts. But even in the complete domain of all real, rational and irrational, imaginary and complex, algebraic and transcendental numbers, there is nothing that would favor an increasing or decreasing series of numbers in the way that a living organism favors for itself the decreasing entropy mode of thermodynamic change.

That there is a developmental sequence common to the historical growth of mathematical knowledge and to the intellectual maturation of the child makes very good sense in terms of the hierarchical structure of nature. The mind necessarily comprises statements (signs and signals) that correspond to the lower Umwelts, which are thus structurally part of the mind, while developmentally, they are its ancestors. The stunning appropriateness of mathematics for the description of the lower Umwelts derives, then, from the fact that through the rules of numbers the mind gives an account of its own lower integrative levels, which it shares with the totality of the atemporal, proto- and eotemporal worlds.

It is appropriate, nevertheless, to take the remarkable correspondence between mathematics and physical reality with many grains of salt. It was during the 3rd century B.C. that Apollonius of Perga wrote his treatise on *Conics;* it was in 1609 that Kepler realized that planetary orbits are conics; it was Newton's universal theory of gravitation that showed why they must be conics, until Einstein's relativity theory showed that they are conics only to first approximation.

ONE of the fundamental doctrines of Pythagoras was the equation of all things and happenings with number. He discovered that the chief musical intervals are expressible in simple numerical ratios. He may have been the inventor of the doctrine which asserts that the sounds given forth by the circular movement of the stars, is a harmony.

In our epoch the foremost arbiter of truth is scientific knowledge which, in its turn, derives its power from the effectiveness of mathematics in natural science. This effectiveness, however, decreases radically as one begins considering the characteristic issues of life, mind, and society.

They said, "You have a blue guitar,
You do not play things as they are."

The man replied, "Things as they are
Are changed upon the blue guitar."

Wallace Stevens
"The Blue Guitar"

Pythagoras. Wood carving from the choir-stalls of the Cathedral of Ulm, made between 1469 and 1471 by Jörg Syrlin the Elder. Courtesy, Deutscher Kunstverlag, München.

Numbers, and what one may do with numbers, make up the language of purest, primordial forms. Their awesome power derives from their appropriateness to the evolutionarily oldest realities (even though the Apollonian knowledge they represent is the latest mode of understanding that became available to man). It follows that as we seek mathematical reports on the biotemporal and nootemporal Umwelts (which the brain shares with all living things and the mind shares with other minds), the messages should become less reliable and less accurate. I am not thinking about unreliability or inaccuracy due to ignorance, but due to the hierarchical nature of causation, and due to the increasingly larger areas of undeterminism on successively higher integrative levels. The rapidly decreasing usefulness of mathematics for the important issues that concern life, ethics, and aesthetics is well known.

Mathematically expressible rigorous rules do underlie biological, mental and societal processes. But whatever is most interesting about life, mind, and society is not in the lower Umwelts of the cells, of the body, or of the brain. For instance, equilibrium statistical theories have been usefully applied to the study of simple social systems. As one example, it was shown that a given population will redistribute itself within the inhabited land, so as to maximize entropy, calculated from the number of microconfigurations (defined, in its turn, as functions of area, people, etc.). The individuals, for the purpose of such calculations must, however, be taken as indistinguishable. If my interest is to predict the likely behavior of hundreds of thousands of people let loose in a certain region, this is the way to make a good guess. But if I wish to know what my uncle Alan decided to do about moving, the mathematical approach is useless. My interest, then, involves the mind and body of a particular man, and those of his features which make him different from, rather than similar to, all other minds and bodies.

Warnings about certain limitations intrinsic in mathematics came to be voiced early in this century. All through the history of mathematical thought it has been imagined that a final form of mathematics ought to be a consistent and complete or completable intellectual structure, from which all other knowledge will follow. This doctrine rests on the belief that mathematics is a deductive discipline subject only to tests of logical consistency, and that logic is a universal science which embraces the principles underlying all others. That all laws of mathematics, including geometry and all abstract forms of algebra, are derivable from and reducible to logic became known as the logistic thesis. By completeness is meant that within the structure of any subdivision of mathematics there are no exioms from which logically contradictory statements may be drawn.

Propositions of this type are examined in proof theory, also known as metamathematics. Its origins are associated with the name of the German mathematician David Hilbert. A fundamental demand of proof theory is that the methods employed in any particular proof must be, in some sense, more primitive (i. e. less complex) hence presumably less dubious, than the methods of the proof whose validity is being tested.

In 1931 the mathematician Kurt Gödel succeeded in showing that mathematics is intrinsically inconsistent and incomplete (in the sense here understood). Even the simplest of formal systems adequate for arithmetic (comprehensive enough to permit addition and multiplication) will permit the formulation of propositions which are (1) self-contradictory in the language of the system though (2) not necesarily so in the language of a more complex system. But rules of a higher system are unspecifiable in the language of the lower system, hence the hierarchy of mathematical languages must be regarded as inductively coupled. In one fell swoop, Gödel also showed that Hilbert's program of proof theory is impossible to fulfill, for as we go from proof to proof, the methods used to prove consistency become more complex, and hence more suspicious.

This whole scheme, known as Gödel's incompleteness theorem, bears a remarkable isomorphism to the theory of time as conflict. On each level of the open-ended hierarchical structure of mathematical knowledge we find self-contradictions which are inexplicable in the language of that level. In the theory of time as conflict we identify conflicts which are unresolvable on their indigenous levels of occurrence. In mathematical knowledge, the self-contradictory sentences may become explicable in the language of the next higher level. In the theory of time as conflict the resolution of the earlier conflict may become possible through the emergence of a new integrative level – with its own unresolvable conflicts. In proof theory subsequent proofs become more complex, and hence more dubious. In the hierarchical structures of temporal Umwelts each world must be described by a more complex language and, on each integrative level, its own language leaves increasingly larger areas unrecognized, and hence undetermined.

I assume that this isomorphism between Gödel's incompleteness theorem and the theory of time as conflict is a witness to their common origins. I propose that the common roots be sought in the structure of the brain, which is known to have retained the hallmarks of its developmental stages. Human language, I said earlier, is a process description of the brain; mathematics is part and parcel of that process description. It is made up of signs that stand for the symbolic transformation of experience by the individual, and by the species. The rules of mathematics, as the rules of logic, are rooted in the capacity of snails, horses, babies, and mathematician-kings to separate the permanent from the contingent in their internal charts.* In the words of J. A. Schossberger,

> the striking congruence between the inner syntax of mathematical methods and their supposedly objective observational counterparts [resides in the fact that] the external sign system of mathematics and symbolic logic are merely precise and increasingly compact restatements of human experiencing. This is so because in contradistinction to objects and lifeless configurations, the symbols given to us by our predecessors and reposited in the environment convey the recorded expression of their creators' experiences and intentions. Mathematical treatises, operation manuals and the like belong to this category of symbolic extrajects.

*Cf. the biological argument which sees language as belonging to the latest stages of speciation of Homo sapiens with the capacity of mathematics, according to Lenneberg, originating as a slight modification of linguistic ability.

If intellectual history is any guide, applied mathematics will tend to progress so as to subsume within its domain such functions as belong to the biotemporal, nootemporal and sociotemporal Umwelts. But in its career, if mathematical knowledge is to become appropriate for the higher integrative levels, it shall have to lose its cherished rigor and, in its higher reaches, become indistinguishable from non-mathematized knowledge. Quantitative tests as proofs of truth will remain tests of limited applicability.

With what we have learned of the morphogenesis of knowledge, of personality, knowledge, and time, and of the power and limitations of number, let us now turn to the issue of social and cultural predispositions toward the quantification of knowledge.

4   The Quantification of Knowledge

The identification of science as the foremost arbiter of truth is, in our epoch, a world-wide article of faith. In the industrialized nations the indentification is explicit; elsewhere it is implicit in the demand for the services of industrial production, the offspring of science. But mathematized science is not a creation of all mankind; its roots are in the Christian West. It may be argued that the scientific and industrial revolutions could have originated elsewhere, but they did not. I shall assume that the historic birth and growth of quantified knowledge in the West is not an accident of fate but the result of certain predispositions and of social and economic conditions which conspired to produce it, and I shall seek to identify them.

I propose that there exists a significant link between individual predispositions toward certain modes of knowledge and collective judgments of truth. This link, I believe, is as ancient as life itself. It goes back to the division of labor, which made it possible for multicellular organisms to secure for their collective selves more favorable conditions of life than would have been

possible for separate cells acting alone. Under selection pressure
the cells came to relinquish their original autonomy, and became
identifiable only through their functions as parts of a larger unit.
Under the pressure of communal demands, and in the service of
their collective future, individuals learned to yield some of their
autonomy, although one may argue that that autonomy was
originally gained through the advantages of belonging to a
community.

Specialized tasks make sense only in terms of the purpose of
the group. They are created under the guidance of knowledge
which itself is collectively generated. Decision as to what the
individual should contribute is guided by consensus, whether it
pertains to rain-making, methods of healing, or rules of multiplication. In a broad sense, these are communal perceptions of what
is true.

As in organic evolution, so in the differentiation of society,
selection pressure can lead to a division of labor only because
there is a spread of given capacities from which selection can be
made. Out of the available variations collective judgment chooses
those that are useful for the community. But unlike in organic
evolution, where the spread of capacities is determined mostly by
random genetic variations, in the cultural process even the
predispositions for certain modes of knowledge are collectively
suggested and reinforced. In this scheme of epiHemology one
should seek the bases of collective judgments of truths through an
understanding of personalities, for personal preferences and judgments of truth are mutually generative.

The world was already divided into large and distinct
groups of civilizations before mathematized knowledge became
an established method of dealing with nature. Our arguments
should include, therefore, a systematic formulation of what "personalities" we could assign to Hindu, Mayan, Egyptian, Babylonian, Greek and Islamic cultures, in the classic sense of Ruth
Benedict, who saw cultures as "personality writ large." Then we
could ask about their levels of mathematical development and ask
why the scientific revolution did not originate in India, or with the

Mayans. But since we cannot carry out such a grand project here, we shall concentrate on a single example: China and the West, for the question of Chinese versus Western science has been examined in detailed comparative studies.

Chinese mathematical preparedness and technology preceded those of the West by many centuries. Since Joseph Needham's monumental work on science and civilization in China, there can be little doubt that until about the end of the 14th century the Chinese showed clear technical superiority to the West. In mathematics, physics, chemistry, biology, astronomy their lead was uneven, but still they excelled in giving an account of the world in rational and systematic ways, or, as Ernst Mach would say at the end of the 19th century, in arranging their experience in economical order. But from the time of Galileo onward, the new experimental philosophy of the West overtook and left behind the levels reached by the natural philosophers of China.

A profound interest in nature was consistent with the perennial philosophy of China which Needham (1966) described as an organic naturalism. While Europeans oscillated "between the heavenly host on the one side and the atoms and void on the other" the Chinese worked out an organic theory of the universe which included nature and man, church and state, and all things past, present, and to come.

Theirs was a civilization interested in the exploration and exploitation of people rather than things. Events were seen to have been connected by causation as well as by simultaneous resonance, familiar in the West through the Jungian idea of synchronicity. The idea of time was rarely considered to be independent of astronomy, religion, and the mores and folkways of society. The affairs of the world and man were understood to comprise classes of compartmentalized cycles. By custom immemorial the women of the Chinese Emperor approached him according to the cycles of the moon because of a belief that the virtue of the offspring conceived related to the moon's phases. In this symphony of life cycles, resonances, male and female principles, "history was the 'queen of the sciences' not theology or metaphysics of any kind, never physics and mathematics."

Since the Chinese had a profound interest in nature, theoretical readiness in mathematics, and a highly developed technology, we may well ask why the scientific revolution took place in the West and not in China?

Needham has sought to find an answer to this question. He compares the social homeostasis and built-in stability of China with the built-in instability of Europe, and assigns the restlessness of the West and the emergence of modern science to such socio-economic factors as, for instance, the struggles of independent city states and infighting aristocrats. He sees modern science as one result of the merchants' demands for exact knowledge, and maintains that the differential development of science will eventually be understood in terms of social and economic causes.

But neither restlessness alone nor any constellation of social and mercantile conditions was likely to have brought about a stress on quantified knowledge, had there not been some ideas and practices abroad which suggested that the way to master nature was through the quantification of lawful change.

In search of such ideas and practices we may take our departure from another work of Needham (1961). In a superbly interesting paper, yet one which, to my knowledge, he had not tied to the question of why science originated in the West, he explored the very different attitudes toward human laws and laws of nature in Chinese and Western thought. In the West, some time between the time of Galen and Ulpian in the second century and the time of Kepler at the turn of the 16th century, the laws of nature, as distinguished from laws applicable to man only, came to be differentiated. The laws of nature came to be conceived of as being just as inviolable as the laws that must control man's behavior, both being ordained by a single, almighty Creator. In China similar ideas could not develop, for such reasons as the universal Chinese distaste for abstract, codified laws, the identification of human law with customs and mores, and the absence of creativity among the attributes of Chinese Supreme Beings.

Max Weber in his celebrated work on Protestant ethics

denies that modern natural science can be understood as the product of material and technical interest alone. An ethic based on religion, he wrote, "places certain psychological sanctions (not of an economic character) on the maintenance of attitudes prescribed by it, sanctions which, as long as the religious belief remains alive are highly effective, and which mere worldly wisdom ... does not have at its disposal."

Indeed, by the time of Kepler we may observe the confluence of certain cultural, religious and psychological practices: (1) the existence of a collective personality which advocates and prefers the use of repression as a mode of dealing with instinct and affects, (2) practices and ideas that not only sanctioned, but praised mathematical exploration as a profitable and noble venture for body and soul, and (3) a new emphasis on the importance of the future well-being of man on earth. These factors form a single syndrome of mutually reinforcing practices and views.

The roots of this syndrome go back perhaps as far as the reevaluation of sexual mores associated with the ministry of St. Paul. Practices contemporary with Christ suggest veneration of the power of sex in the male but condemnation of it in the female. The thrust of Roman Christianity has been the degrading of the feminine in life by upgrading, perhaps from the 4th century on, the virginity of Mary. In terms of the two forms of genital knowledge which we identified earlier, these ideas favored aggressive penetration, in contrast to the Chinese preference for organic internalization of knowledge.*

The daily routine of the Christian in general, and the Protestant in particular, emphasized guilt and fostered anxiety, by insisting on the repression of instinctual drives. But repression of instinctual drives is known to favor a shift from genital knowledge which explores people (such as the heavenly bodies of the Chinese

---

*It must be remembered that modes of genital knowledge are not limited to members of the respective sex, and that preferences for the female type of genital knowledge are not to be mistaken for a statement on the social status of women.

Emperor's wives) to ocular knowledge that explores forms and objects in space and seeks order and balance (as in the heavenly harmonies of planetary bodies). Sublimation was encouraged through strict control of the self and the environment and through parsimony of feelings and self expression. Mathematics is the ideal of parsimonious thought, and the absence of values in mathematical symbolism makes it a safe haven against the upswelling of humane concerns. Thus, mathematics suggests itself as the ideal means whereby the discipline of the Divine may be appreciated. Not incidentally, the application of number also brought more profit to the merchant, a fact which surely offered further proof of the divine origins of quantified knowledge. Mathematized knowledge permitted the systematic testing of abstract hypotheses against natural phenomena; affirmative results then amounted to a continuous reconfirmation of the timeless power of the Christian God. For the mercantile economy of the West, for the clockwork universe of Newton's epoch, as for Plato, God was a mathematician – although for different reasons.

Beyond these psychological elements there were also theoretical predispositions in Christianity toward the quantification of knowledge. "The Christian knows," wrote Kepler in 1619, "that the mathematized principles according to which the corporeal world was to be created are coeternal with God." But how does the Christian know it?

Alexander Kojève has suggested that to embed precise mathematical relations in nature demands the possibility that matter harbor perfection and, in Christian thought, in contradistinction to the idealism of the Greeks, matter had done so in the incarnation of Christ. Kojève's generalization cannot be strictly maintained, but it must be admitted that such views as he stresses must have served as sources of encouragement.

I would like to point to a less subtle and philosophically more primitive issue. How could the faithful reconcile their awe for the timeless foundations of reality with the daily emphasis, peculiar to a mercantile economy, on the importance of benefiting from the passing time through diligent work? The time and the timeless were joined in their world-view, ready made as it were, by the dogma of resurrection. Everywhere in Christendom, repeated day after day, alone and severally, the faithful declared: "Credo in unum Deum ... remissionem peccatorum, carnis resurrectionem, vitam aeternam." It was a package deal: eternal, timeless life and the guilt of man existing in time, were joined in potential resurrection for all. This theology granted religious authority to linkages between time and the timeless, and what better linkage is there than that between the timeless rules of mathematics and the temporal world of sensual reality?

Yet another contributor to the theoretical predisposition toward the quantification of knowledge may be found in the Judeo-Christian Heilsgeschichte. Its central message is a promise of a better world in the future. Not unlike salvation history mathematized knowledge also possesses powers of prediction, hence suggests the possibility of a controllable future. Technology, an offspring of science, explicitly offers the potentiality of a future better than the present. The coincidence between the promises of Heilsgeschichte and of mathematized knowledge can be used to argue their mutual validity, especially if one forgets that the praise of quantified knowledge is itself a derivative of salvation history. In a cultural milieu which does not allow for the possibility of long-range improvement in the state of man, the teachings in which promises of improvement are implicit are not likely to be judged as acceptable, and the mathematized paradigm would remain unpopular. From the psychological point of view, emphasis on the future directs libidinal energies toward distant goals and helps sublimate instinctual drives toward increased possession of goods and control of time budgets. In the stable world of China, salvation history was ordinarily not called for; security was vested in the emperor and not in abstract ideas.

What I have said so far confirms the suspicion of Max Weber, first formulated in 1904. The origins of natural science are to be sought, he wrote, in the decided propensity of Protestant asceticism for empiricism, rationalized on a mathematical basis. "The favourite science of all Puritan, Baptist, or Pietist Christianity was thus physics, and next to it all those other natural sciences which used similar methods ...."

Nothing that I have said so far should be construed as a claim that mathematical physics emerged, somehow, from nowhere, due to the magic of certain practices and attitudes. Even Pythagoreanism is a mathematization of nature, and Ptolemy's numerical demonstrations in the *Almagest* are as impressive today as they must have been during the second century. But the great importance of philosophical, mystical and religious motives cannot be neglected without making the whole enterprise unintelligible.

Beyond the broad identification of the historical Western preference for the penetrative over the internalized mode of genital knowledge, we may identify other related issues. For instance, for Plato, sight was the keenest and most wonderful sense. With St. Paul, Christians on earth see things as though through a glass darkly, but after death "face to face." The clear and unclouded mind – the ocular mode – is surely the preferred Cartesian method of knowledge. In contrast, some other cultures prefer oral knowledge and emphasize hearing over sight. In India chanting is still the most direct way to the truth of the timeless. Islam is a religion of the spoken rather than of the written word. And the Hebrew prophets, those giants in search of identity, heard rather than read the truth. God spoke to Moses and Moses spoke to the children of Israel: "Hear, O Israel; The Lord our God, the Lord is one."

With the rich variation in favored methods of action and thought, it is not surprising that even mathematics, the most universal of all languages, is spoken in many idioms.

Enough has been said to suggest that a configuration of practices and views of Western Christendom, around the time of Galileo and Kepler, provided an atmosphere that favored the emergence of the scientific way of looking at the world. These conditions were most probably necessary, though perhaps not sufficient, to generate that historic motion which, in our own epoch, is made manifest as the identification of truth with what is scientific.

We found earlier that quantification is most appropriate for

·our knowledge of the lower integrative levels. It follows that insistence on mathematical precision as a necessary corollary of truth bestows upon the individual certain psychological benefits. Specifically, it relieves the individual mind by drawing the mind's attention away from the conflicts of the nootemporal Umwelt. Unlike good poetry which shakes the soul by its ego-disturbing truths, the power of logic and mathematics resides in their safety, in their lower temporalities or (in unexamined language) in their timelessness.

Albert Einstein (1949) described once how he labored to remove himself from a merely personal world of wishes, hopes, and primitive feelings. The contemplation of a man-independent, extra-personal world beckoned to him like a liberation. "In a man of my type," he wrote, "the major interest disengages itself to a far-reaching degree from the momentary and the merely personal and turns toward the striving for a mental grasp of things."

5   The Fellowship of Science, Technology, and Truth

Without any doubt, mathematized knowledge is the first truly universal language (although spoken, as it were, in different idioms); the scientific method is the first truly universal ritual that applies equally to all men without regard to race, sex, color, or national origin. All major religions have attempted to gain for themselves such a universal acceptance by word and by sword, as have many civilizations. Yet none has succeeded, nor is any likely to succeed, because the heterogeneity of the biological, noetic, and cultural forms of man works against them. Quantified knowledge has a powerful advantage over religions and social doctrines: it derives from, and applies to, those lower integrative levels which are shared by all matter: thinking, alive, or inanimate.

Plato's geometrician God constructed his changeless order; Leibniz' architect God erected a changeless structure. In our epoch the reliability implicit in the laws of nature, identified mainly with the laws of mathematical physics, has become the closest thing to the absolute and the divine. It is upon the predictability of these laws that the imposing structures of technology and industry were built. We shall examine briefly the intellectual history of technology as a mode of knowledge, and tie our findings to a critique of the symbiosis of science, technology, and truth.

In the history of technology there is a clear continuity from pulleys and levers (useful because they function as expected), to current-carrying conductors that move in magnetic fields as expected (and make motors), to machines that light up with a brilliance lighter than a thousand suns as expected (for they are atomic bombs). Since the Renaissance, verifieable truth has been increasingly identified with such propositions that lend themselves to the test of predictability, and predictability with whatever ideas and propositions that can be "made to work." This practice gives science a clear advantage as the final judge of truth, over ethical or aesthetic propositions. For the sources of the common cause which thus binds science, technology, and truth, we shall look into certain individual and collective practices and preferences of the Christian West.

We distinguished earlier between the masculine and feminine forms of genital knowledge. Because of the hermaphroditic freedom of the mind, a healthy and mature individual is usually free to prefer either mode, but the advantages of free will should not obliterate the polarized origins of what I labeled masculine and feminine modes of inquiry. Let us recall Erikson's experiments discussed earlier in this chapter. They demonstrated one way in which human knowledge is born: biological forms and functions, projected across the biotemporal/nootemporal interface, become paradigms of human behavior. Morphological differences between sex organs are metamorphosed into differences in life styles and, so I argued, even into distinct ways of thought.

The distinguished British biologist Brian Goodwin, has put forth some interesting ideas in embryogenesis. They are homologous with those of Erikson. Goodwin argues that the conceptual steps we must use to explain how an organism is generated from an egg are the same as the ones we need to explain how useful representations of the experienced world are generated by the mind. There is, he asserts, a continuity between embryogenesis and behavior. Chemical substances synthesized by embryo cells display context-sensitive effects, with the embryo itself creating the different contexts for the behavior of its different parts. Within a spatially ordered set of cellular activities, specialized cells amount to useful representation of some aspect of the experienced world. They are, one might say, cognitive systems appropriate to the cellular world, just as our thoughts are appropriate to the world of the mind. In our terminology, both Erikson and Goodwin deal with the projection of forms and functions across the biotemporal-nootemporal interface.

We remember now the Western bias for ocular, as against aural knowledge. Ocular preferences seem to be affine to the masculine-genital mode more than to its feminine-genital counterpart, possibly because aggressive penetration is naturally directed toward objects in visible space, whereas functions that enclose and protect are more likely to be veiled to sight. If we put all these together, the following structure of thought emerges.

Under the heading of masculine-genital knowledge, with preference for ocular exploration, we may class behavior that centers on building and on being fond of things that work, whether constructively or destructively. Concern with the "outer space" would include such disciplines as astronomy, chemistry, and the life and mind of others. Because of its ocular character, it would include preference for the visible, predictable, analytical, balanced, mathematical, the inorganic. Under the heading of feminine-genital knowledge, with preference for aural exploration, we may class behavior that centers on enclosing, and on being fond of things that live, whether constructively or destructively. Concern with the "inner space" would include creative functions which are normally hidden, and interest in life and mind as experiences rather than as external processes. Because of its affinity to the aural, we would also class under feminine-genital all knowing that is usually gained through "voices," and also whatever relates to the contingent, to the organic, to feeling, to rhythm, to wholeness. It is the privilege and the curse of many mature minds, and a hallmark of many highly creative people, that they combine the expansive-eruptive and the ocular with the in-gathering, rising-and-falling, and aural modes of knowledge. Recently I described this cohabitation of personalities as that of "passion and knowledge," a contemporary form of the ancient yin and yang principle, and an unresolvable conflict of the mind.

Let us shift our attention now to the knower: never a passive recipient of some absolute truth, but a creative participant in establishing that truth. There is no such thing as immaculate perception, but only passionate knowledge. The corollary of this claim is that truth itself is Umwelt-specific, an idea necessary for the self-consistency of the theory of time as conflict.

With the deep-seated conviction that quantification is tantamount to science and that science is the final judge of truth, our fields of knowledge strive to imitate the rigor of mathematical physics, quite oblivious to the degradation of Umwelts which must accompany the quantification process. Because of the hierarchical structure of languages and the fact that, on higher integrative levels, larger regions of phenomena are unacknowledged, hence undetermined, insistence on increasing precision and generality leads only to the understanding of increasingly lower and narrower segments of nature.

Compare quantum theory with chemistry used by working chemists. The general laws of quantum theory can be stated with very great precision (albeit, probabilistically) and possess great generality whereas the formulas and tables used by working chemists and engineers comprise mostly non-universal instructions. In principle, quantum theory underlies everything, whereas practical chemistry deals only with the making of paints, medicines or tires, and has none of the elegance of particle physics. But there is no known way whereby we could proceed from the theory of hydrogen atoms and arrive at the rules of curing ham.

In physics we observe a successively increasing universality and a simultaneous regression along the Umwelts. For instance, in kinematics the Galilean transformations are invariant for stationary observers only; the Lorentz transformations are invariant for stationary and translating observers; general relativistic transformations are invariant for stationary, translating, and accelerating frames. But whereas the Galilean transformations are founded on the experience and idea of a local eotemporal world of bodies, the power of relativistic transformations is rooted in the atemporal character of light, and relativistic transformations have their major significance in the atemporal and prototemporal worlds.

In biology, a degree of realization of the hierarchical character of lawfulness has come to be admitted. Thus, evolution by natural selection is not a precise statement but a general principle which is mostly explanatory and not predictive. But the major body of biological inquiry is going the other way: toward quantitative biology. This trend informs not only the working habits of the best of biologists but also biases their estimates of the future course of human knowledge and the fate of mankind. Thus, for instance, E. O. Wilson holds that the ultimate goal of sociobiology is a stochiometry of social evolution. When perfected, the respective laws will permit the quantitative prediction of the qualities of social organization, including division of labor and time budgets. These will be calculated from our knowledge of the prime movers of social evolution, which are phylogenetic inertia and ecological pressure. Eventually

> the humanities and social sciences [will] shrink to specialized branches of biology; history, biography, and fiction are the research protocols of human ethology; and anthropology and sociology together constitute the sociobiology of a single primate species.

There are, in his view, certain dangers.

> If the decision is taken to mold cultures to fit the requirements of the ecological steady state, some behaviors can be altered experientially without emotional damage or loss in creativity. Others cannot. [Finally, if individuals will be relieved by social design, of the] stresses and conflicts that once gave the destructive phenotype their Darwinian edge, the other phenotypes might dwindle with them. In this, the ultimate genetic sense, social control would rob man of his humanity.

The dangers perceived by Wilson are very real: if the unresolvable conflicts of the mind, which are the sources of man's creative as well as destructive capacities, are removed by collapse, the nootemporal integrative level and its source, the self, vanish.

This presumptive decline is greatly facilitated by the error of misplaced precision, exemplified by the first quotation. The mad rush for accuracy, and the identification of acceptable truth with quantification, will surely lead the science of life, mind, and society down the garden path. By the time biology, psychology and sociology become as rigorous as mathematical physics, they shall have left by the wayside whatever is unique and interesting about man: his life and death as an individual, his hierarchy of unresolvable conflicts, his insistence on the good, his preference for evil, his desire for the beautiful.*

If quantifiability remains the leading criterion for truth, then mathematized science is well qualified to be the final judge. But the common cause between science and truth is a defensible proposition only if technology, proud infant of science, proves itself as naturally assisting rather than opposing ethical and aesthetic judgements. Plato, source and godfather of the incorruptible geometry of modern science, warned us that the true, to be true, must be illumined by the good. But boys will be boys, and judging from the crises of our epoch, the holding up of the scientific method as the model of tests for truth does injustice to man – and to science.

Lewis Mumford, able observer of the contemporary scene, complained recently about the pious beliefs of the futurologists. That limitless progress

> carried with it the possibility of an equally indefinite human regression did not, as late as the nineteen-sixties, occur to the

---

*Belief in reductionism, wrote W.H. Thorpe (1975), subconscious though it may be, has an effect on the sanity of modern man. Reductionism today is a mask for nihilism, which in its contemporary form is not nothingness, but "nothing butness."

faithful disciples of Marshall McLuhan, Daniel Bell, or Arthur Clarke – those giant minds whose private dreams all too quickly turned into public nightmares .... Did not the president of a great university the other day publicly salute Buckminster Fuller, that interminable tape recorder of "salvation by technology" as another Leonardo da Vinci, another Freud, another Einstein?

Only mammalian tenderness and human love have saved mankind from the demented gods that rise up from the unconscious when man cuts himself off from the cosmic and earthly sources of his life.

In our search for truth, it would appear rational to include ways of knowing which need be neither precise nor quantifiable, and whose protocols do not demand reproducibility. Epistemology ought to be removed from its present position as a mode of criticism within the narrow confines of the so-called philosophy of science, and replaced in the position whence it came: a theory of knowledge, in general. The call for such a shift is one manifestation of the general need to enlarge our modes of inquiry so that they include methods and interests that are better identified with the female, than with the male principle. New categories of lawfulness are needed to accommodate those departments of knowledge which are complementary to the precise and quantitative approaches of exact science. This claim is itself extrascientific, hence unprovable with quantitative rigor.

Truth, then, is only one class of knowledge and its power is subject to challenge and change. If truth is identified with what is quantifiable, precise, and predictable, its human value becomes severely limited, for it can pertain in most cases only to the lower integrative levels of nature.

X   Society and the Good

IDEAS OF GOOD AND EVIL express communal preferences for praiseworthy conduct. Although, with such a general formulation, good and evil could be defined for social animals in general, our concern is man. For him, the instructions of his collectives take the form of selection rules, designed to help him choose a path of action among the many that his imagination can conjure up. The foremost arbiters of preferred behavior have been religious and political visions, so it is these that we shall consider first. Our findings will suggest that the evolutionary office of morals is the creation and maintenance of disequilibria, or conflicts of certain type, and the control of their resolutions. I shall conclude by using these ideas, and try to elucidate somewhat the crisis of meta-stability that characterizes the world community of our age.

1   Good, Evil, and the Religious Vision

With the broad evolutionary definition of good and evil just proposed, the decision where to start our inquiry is a question of expository technique. One could begin with mountain brooks, for they run along gravitational gradients so as to return to their own kind, as Aristotle might have had it. Such a return, one might say, would be praiseworthy moral conduct for water, provided it had the choice of not flowing along the gravitational gradient. But it does not have that choice; hence the path taken by the brook remains a statement of truth and as such, of concern only to science; the issue cannot involve questions of behavior.

Perhaps a billion years after water precipitated on the hot surface of the earth, the simplest animal societies appeared, such as that represented in our own epoch by the cellular slime mold. The life cycle of this mold includes free living and colonial stages. In the free living mode the organisms act as independent amoebas. As the population density increases, the separate amoebas are called together by a chemical signal released periodically from a pacemaker cell. Some amoebas begin to congregate, climb upon one another, form a mound; the individuals as individuals vanish in the process of forming a pith whose parts are now intimately joined by chemical means of communication. The chemical signals diffuse beyond the limits of the stalk and induce uncommitted amoebas to follow the concentration gradient toward its source and thus join the colony. The collectively preferred behavior for a free amoeba resides in joining the colony so that the life of the species may continue.

Three billion years after the simplest animal societies began to function, temples and cities came to play roles in archaic societies, analogous to those of slime mold pacemakers and stalks. They were navels, axes, and centers of the world toward which the individual tended to gravitate; in myth and religion they were often described as earthly embodiments of heavenly archetypes, usually located on a real, or imaginary, hill. Praiseworthy behavior for the individual consisted in approaching such centers with respect and concurring in their demands. Thus it happened that "Joseph also went up from Galilee ... unto the city of David which is called Bethlehem," following the gradient of his mission, "to be taxed with Mary his espoused wife, who was great with child."

Turning from phylogeny to ontogeny, in the life of the infant the mother's appearance, when needed, is probably the earliest experience that can give meaning to what later become ideas of trust, fulfillment, and the good. It has been conjectured that the mother's breast is the early paradigm of good objects; the capacity to anticipate its appearance in imagination lessens the traumatic experience of helplessness that must accompany its delayed coming. But imagination cannot, in the long run, replace reality; if relief does not come the hunger increases and the good object changes into a bad one, hope gives way to anxiety, to fear, and to impotent rage, in which the utmost good turns into the utmost evil. The infant cannot articulate his feelings and does not yet know noetic time. But some years later, as a dying man, he reencounters the same duality in mortal terror. Life, the utmost good, gives way to life, the utmost evil, and he has a question: "Eli, Eli, lama sabachtani? That is to say, My God, my God, why has thou forsaken me?" But he has no answer, only an unresolvable conflict in which futurity and pastness collapse into an unbearable, eternal present, and his body returns to dust. Thus, the knowledge of good and evil accompanies man from cradle to grave, and has been doing so for ten million years or more. During this period good and evil have taken on the innumerable forms that only the broad spectrum of man's needs and his imagination can create and accommodate.

Ancient art and artifacts suggest an archaic belief that certain deeds in the present might sway fate and improve inauspicious conditions in an uncertain future. But this future does not stop with the death of a person: burying weapons, tools, ornaments and food with the dead, already in the Paleolithic era, indicated a firm belief in post-mortem survival. It seems that the discovery of personal death is as old as the rejection of its finality; the mind, as we learned, resents any interference in its affairs by the mortal body. The presumed history of the self is unrestrained by the limitations of the soma. I shall argue that the roots of religious beliefs may be found in this peculiar freedom of the self, in combination with a realization, already by archaic man, that the future, in general, is loaded with promises of good and evil.

A steadily enlarging temporal perspective may be detected in the progression from the rituals designed to influence tomorrow's hunt, to prayers for liberation from the Waters of Babylon, to supplication for freedom from human bondage in general. As socially and individually desirable goals came to be extend to longer futures, the symbolic transformation of experience represented by religious rituals, designed to facilitate the reaching of those goals, had to become increasingly abstract, because, in fact, the more distant was the goal the more uncertain was the future. With the increasing autonomy of the mind, the goals themselves often changed from physical things and conditions here and now, or tomorrow, or even right after death, to things and conditions in a (practically) unreachable elsewhere. The good and evil of daily experience came to be reified and sublimated into ethical ideals. Such ideas were integral parts of universal cosmologies and cosmogenies, which spoke about the origins and nature of the world and the ways man ought to guide his daily life. Let us consider some examples.

○ The founders of Buddhism dwelt on the ills which beset the individual's existence on earth and sought to advocate a system of conduct that would lead to deliverance from suffering. It is difficult to identify a common essence among the many beliefs and practices of Buddhism, because during twenty-six centuries of its existence pronouncements have been made on almost every conceivable aspect of life and the world. Collectively, these utterances are interpretations of human life and destiny. They are designed to win release from suffering by a denial of the reality of what, in the theory of time as conflict, are seen as the unresolvable conflicts of the mind. Accordingly, approved and praiseworthy behavior consists of such thoughts and actions as are in harmony with the assumed unreality of the self. The conflict is solved by regression.

○ Hindu interpretations of human nature and destiny have also been many. The synthesis now known as Hinduism stems from cults that date to the Vedic period, roughly 1500–600 B.C. Such hallmarks of Indian thought as the unwillingness to credit history with any metaphysical content, emphasis on perfect beginnings and subsequent deterioration, or on the eternal repetition of a fundamental cosmic rhythm are closely bound to the world-view of cyclic time. Its representation is the metaphor of the "sorrowful, weary wheel." Value judgments derived from such cyclic world-views, combined with the insecurities that have been rampant on the Indian subcontinent for millenia, suggest suffering as the only reality. Man, stooped under the weight of daily care, can escape his miseries only by escaping from temporality, by transcending the human predicament of existing in time. In the words of S.G.F. Brandon (1962), salvation lies in flight from a universe "in which the twin process of creation and destruction pursues inexorably its awful way; salvation ... lies in the virtual oblivion of ecstatic union with [the] divine Lord."

The cyclic stratification of time has its analogy in the stratification of the Hindu social system. The most important order to preserve is that of society; after society, that of the family, itself also stratified. Consistently, social Hindu ideals of rightousness, material advantage, and merit have been subordinated to salvation, pursued by ascetism. Ethical behavior is seen as that which promotes the resolution of the unresolvable conflicts of the mind, by regression.

○ As already emphasized, the historic Chinese view sees man as one with the cosmic process. According to Taoism, the most ancient and fundamental principle in Chinese culture, well-being consists in integrating the individual with nature; hence he must understand Tao, the way of man and the world, and conform his life thereto. Unlike in the case of Hinduism, the Chinese preoccupation with the cyclicity of time did not, however, lead to a prevailing belief in the unreality of time.

Philosophical Taoism advocates the taking of no "unnatural action"; religious Taoism seeks earthly blessings and advocates filial piety, loyalty, love, harmony. To secure the desired harmony between the individual and the universe, Taoism is predisposed toward mysticism, while also demanding a high standard of intellectual endeavor. In the view of the celebrated Sinologist Marcel Granet, the Chinese summum bonum is a long life in this world, during which, by carefully conforming to his own peculiar way, the individual promotes the universal Tao, the eternal cosmic rhythm of Yin and Yang. Praiseworthy conduct, then, resides in integrated tranquility. To what extent the profound social changes in contemporary China, advocating in word and deed the unremitting drive for achievement, will influence the traditional Chinese evaluation of man's position in the universe, is yet to be seen.

○ For the ancient Iranians time was deified as Zurvan, progenitor of the two opposing cosmic principles of good and evil, creation and destruction. In Mithraism (the worship of the Indo-

Iranian god of light, Mithra), the interplay of creation and decay is combined with appreciation of a savior of hope, a heroic leader who offers deliverance. Mithraism has spread from ancient Persia to Rome and the West. To what extent, if at all, its interpretation of the human experience as an extended conflict is an ancestor of the Christian view, is debatable. The Hebrew and Aramaic manuscripts popularly known as the Dead Sea Scrolls, believed to have originated around the time of Christ, strongly stressed the need for awareness of the struggle of light and darkness.

In this feature, the teachings of the Scrolls resemble yet another dualism, one of Babylonian origins: Manichaeism. The system of Mani, "the apostle of God," originated some time during the 3rd century. It saw the world as the conflict of two independent and coeternal principles: Light and Darkness. After a long and complex struggle, the Powers of Light provided a framework for salvation through the revelations of Buddha, Zoroaster, and Jesus. The earthly path to the Light was rigorous: the Manichaeian ethics was ascetic, for the demands of the body feed but the powers of darkness, and only the dictates of the soul please the Light.

○ The authors of the Dead Sea Scrolls, as well as Mani, may be considered forerunners of Gnosticism, a religious movement of late antiquity. It is generally believed that the development of early Christian doctrine was to a large extent a reaction against Gnosticism. Certain gnostic ideals have proved enduring. The consubstantiality of the soul with the Godhead, hence the mystery of the self; and the idea that evil is due to a break with God, survived through such idealist thinkers as Goethe, Novalis, and Hegel. The development of modern Gnosticism with its emphasis on equating the actualization of the person with the actualization of God, is an obvious reaction to catechized Christianity. When Nikos Kazantzakis sees a man as the Savior of God who "is not Almighty, but struggles, for he is in peril every moment; he trembles and stumbles in every living thing ... is defeated incessantly, but rises again," or, when the theory of time as conflict sees the dichotomy of good and evil as level-specific and itself evolv-

ing, they articulate in different languages the unresolvable conflicts that determine the temporality of man.

Early in this section we left Joseph and Mary on their way
to the tax collector in Bethlehem. Their journey signifies the
crucial epoch of Christian Heilsgeschichte (salvation history), for
in that view, it is the life of Christ that separates the informed
from the uninformed portions of history. For two millenia the
interpretations of the life and death of Christ have given direction
to the preferred Western mode of personal and collective conduct.
Because of the far reaching social changes that Christian behavior
has facilitated in Europe, such as the scientific and industrial
revolution, salvation history has also profoundly influenced the
ethical stance of all people on earth.

The Christ story, from which Christian ethics unfolds, is a
remarkably stable constellation of archetypes. It is a lasting drama
of guilt and punishment where the issue of who is guilty and who
is being punished is rather ambiguous. The life of Christ is only a
prologue to His resurection and to the subsequent dogma of
resurection, which assured the faithful of a direct link between
time and the timeless. The individual, laboring for the post-
mortem survival of his soul, could identify this struggle with
Christian identity; the earthly salvation of the Hebrew covenant
was replaced by the heavenly beatitudes, and in due course,
through the intellectual edifice of the Church of Rome, the
beatitudes themselves became the ethical paradigm of the West. It
was with the evil deed of Adam that man learned of sex and time,
and it was with the good deeds of the Christian that he could
return to the sexless and timeless Paradise of his childhood.

A break with this heaven-orientedness came with the
Reformation or, more precisely, with the post-Reformation refor-
mulation of the principles that bind morality to Christian faith. In
the Roman church the preferred way of coping with affects has
been the sublimation of drives, for the benefit of the City of God.
This was expressed in such customs as the admiration of the
Virgin Mary, and in concurring with the demands of Rome. In
contrast, the Lutheran fear of the Devil, invariably a male

chauvinist character, favored sublimation for the benefit of the City on Earth. The soft qualities of the beatitudes came to be replaced by the harsh ideals of Protestant conduct, useful in building the towns of wealthy merchants, who offered their good deeds in barter with the devil within. The Lutheran parsimony of time and emotion, which figures so profoundly in the origins of mathematized science, is basically an ethical stance. Thus, the abstract future-directedness of early Christianity came to be focused upon our future on earth and produced, in due course, its secular child, the idea of progress. The Enlightenment rejected the supernatural trimming but retained the idea of progress as identically equivalent to the good.

The mutual dependence of ethical and cosmological teachings is profound. Rituals, religious teachings, and myths traditionally subsumed ideas of truth as well as instructions for conduct. Since the same teachings usually included claims of foreknowledge, tying ethics to the nature of the universe and to time was an easier task for religious ethics than it was for rational ethics. The secularized idea of progress lost its skyhook. The parsimony of emotive depth of scientific truth precludes the delivery of a code of conduct with convictions both deep and convincing. Hence, we are confronted with an unending stream of secular redemptive measures: political systems, magic pills and potions, and the exotic lure of modes of life that have already been examined long ago and found wanting.

In all religions and cosmologies of which I am aware, there is an asymmetry between attitudes towards states that preceed birth (or the creation of the universe) and states that follow death (or the end of the universe). Non-being before birth is usually regarded with indifference or curiosity, non-being after death with awe, subserviance or fear, or even perhaps with expectation filled with intense joy. I have never come upon a poem or a treatise inspired by non-being before conception, but there are libraries filled with writings about post-mortem existence or non-existence. For various psychological reasons, memories of the destructive aspects of time tend to be more overwhelming than those of its creative aspects. The works of the Evil One are more frightening than the works of the Good One are encouraging; hence the need for reassurance: "though I walk through the valley of the shadow of death, I will fear no evil: for thou art with me."

There is a distinct difference in literary flavor between legends of world creation and those of world destruction, even in religious philosophies which conceive of history as an eternal return, and regardless of whether the golden age is at the beginning or at the ending. Whereas myths of creation tend to appeal to the intellect and propose to explain the things and processes of the present, visions of endings are usually moralistic, restrictive, and strongly suggestive of preferred behavior. To the extent that religious teachings about man and the world influence behavior, they do so by eschatologies rather than by cosmogonies.

Interpretations of past events have been argued using both secular and religious knowledge, but there is no significant non-religious body of inquiry into questions of the eschaton, that is, the last things. Yet it is thoughts of endings that tend to set the patterns of preferred behavior, not thoughts of origins. In the Christian West, the final Apocalypse is carried over into modes of expressing value judgments in everyday language. Consider that "right" and "left" in English, and in many other Western and mid-Eastern tongues, are much richer in meaning than is necessary for telling people which way to turn at the street corner. "Right" is also "correct;" in politics it means the familiar; in law "right makes might." The left is "sinister," from the Latin "sinister" which means perverse, unfavorable, inauspicious, unlucky; a left-handed compliment is an insult. Christ, having ascended into heaven "sitteth on the right hand of the Father." On Judgment Day He shall set the good ones on his right hand, the evil ones on the left; the ones on the right shall inherit the Kingdom whereas the ones on the left will be sent to everlasting fire. Life well lived is rightous until cut short by sinister death.

In those religions that admit a dualism in the nature of the world, the struggle of good and evil usually causes frustration to the Divinity. God witnesses the struggles and setbacks in the striving for individual perfection (as in Buddhism), or the apocalyptic conflict of good and evil (as in Christianity), or the cosmic fight of light and darkness (as in Zoroastrianism) but He, She or It knows that when time from time shall set the world free, the good, the light, or the perfect will win. In humans, frustration tolerance is a measure of maturity; on this scale God or the gods are the most mature beings, capable of waiting for the longest time. Religions, no less than the sciences, project the personality traits of their creators upon the heavenly (or astral geometric) screens. Instead of an old man with a clock as in scientific cosmologies, here we have an old man with a mortal soul. In those religions that teach specific cosmic metaphysics, the greatest good usually resides in the imitation of, or approach to, the godhead, and in being on that side of the struggle which will eventually be victorious. Conduct that promotes such an anticipated victory, or at least maintains a status quo, is usually judged as praiseworthy, its opposite as undesirable.

But if praiseworthy conduct is somehow tied to the fate of a select group, small or large, then the appropriate instructions can best be given in the language of that group. We found earlier that languages appropriate to a higher integrative level are apt to be unintelligible in terms of lower order languages. Since ethical instructions are societal in their origins, their appropriate language is that of the societal Umwelt. It is, therefore, not surprising that moral guidance often has the aura of having come from "beyond and above." This from-beyondness is implicit in all great religions and has been explicitly claimed by the genuine religious leaders of mankind. As we shall see later, the hierarchical structure of integrative levels precludes the formulation of any completely satisfactory method whereby the value of ethical instructions may be tested on the level of the individual mind. It is therefore not surprising that the from-beyondness has also been abused by innumerable minor lights whose obscurantism is supposed to be regarded as bona fide proof of access to higher truths.

By the middle part of this century scientific knowledge

opened up a world of vast dimensions and curious lawfulness. In that world the Hebrew Heilsgeschichte and the Christian soteriology, which together suggested and justified Christian ethics, became untenable. But nothing replaced the universal teachings of myth and religion; man on earth lost his cosmic anchorage except for the obscure linkage through the "old man with the clock." The task of formulating norms of praiseworthy conduct was passed on to science as the sole authority of truth. But science, especially exact science, deals with the lower temporalities only; hence it is ill-equipped to say anything important about good and evil. The confusion and malaise of modern man as regards ethics has followed.

We are going to backtrack now to follow up on a development which parallels, in the West, the historical development of religions: the birth and growth of the secular evaluation of man and society.

2   Clocks and the Political Vision

The concept of the good has been tied, more often than not, to the interest of the polis; hence as the city changed its position among the affairs of God and man, so did the idea of the good in politics. Plato's city is akin to the Greek cosmos; Augustine's is a model of the Christian heaven; Machievelli's city is Florence; Hegel's is tied to the spirit; the city of Marx is a factory owned by the proletariat; the city of sociobiology is a product of social organization, which is "the class of phenotypes furthest removed from the genes."

The turning point in this story cannot be precisely identified, but it may be represented by the early synthesis of the medieval heaven and the Renaissance earth, such as in the thought of Nicholas of Cusa. Ideas inherited from Greeks, Jews, and St. Paul began to merge in a productive symbiosis, and the political and religious ethics of the West began to mix. We shall now consider the historical role of a metaphysical instrument which is a good symbol of what modern man regards as praise-worthy conduct.

The clock began its Western career, perhaps as early as the 14th century, as regulator of monastic life, when, in the words of Lewis Mumford (1963), it "helped give human enterprise the regular collective beat and rhythm of the machine." With the advance of the industrial revolution which it helped to create, it first became a bourgeois, then a Communist, ideal: a symbol of perfection which other machines were to imitate. Man was to adjust his own conduct so as to accommodate this machine. As objects both practical and symbolic, clocks, watches and calendars came to span the distance between, and mix religious and secular traditions. In our own epoch, the clock has become the most important single instrument in the collective enterprise that forces the environment to adapt to man, rather than vice versa. It was not until the transient, romantic revolutions of the second half of the 20th century West that voices have been raised against the clock as a model of desirable conduct.

But let us turn our gaze back to the age of Reformation. The patristic and Medieval ideal of conduct, directed to ethereal, mystical, and distant goals associated with the beatitudes, were replaced by ideals of conduct in which the waste of time, right here and now, was the first and deadliest of sins. Human life was judged too short and too precious to be used for anything but diligent work, so as to assure one's election for sainthood. The parsimony of time and emotion that informed the behavior of the Lutheran merchant suggests an ethical and political stance, as well as a religious one. Whereas medieval man could be deprived of his eternity, the wealthy merchant could be deprived of his present. The "now" of industrialized man, progeny of the diligent merchant, may be saved or wasted, owned or stolen, bought or sold. The worker's labor and his individual destiny in time have come to be seen as two distinct issues.

While the captains of industry organized their armies of workers, the ideology of the industrialized West – captain, worker, and all – became identified with the idea of progress. The practical corollary of this idea is the power of industrial productivity with its promise of creating heaven on earth. Thus Christian salvation history, much changed in form but not in message, came to inform the aspiration of most people on earth; the desirability of *imitatio Christi* gave rise to a secular twin, the *imitatio machinae,* consequence of the admiration of productivity.

But the Umwelts of machines are those of the physical world, characterized in the macroscopic domain by deterministic causation. A good device, whether a clock, radio, computer, bomb, or artificial heart, should function predictably. And predictability, contrasted with the uncertainty of free choice, has a powerful appeal. The machine not only holds the promise of a richer world, but it is also a comfortable companion. As one consequence of its advent, temporalities appropriate for technology saturate the life and bias the views of technological man. The Umwelts of machines (those "circumscribed portions of their environments that are meaningful and effective" for them) are inappropriate even for chickens, let alone for people. Devices of all sizes, functions, and shapes seem to talk in human metaphors: the channels of communication between man and his devices must be a language intelligible to man. Those are our words, but through them, the electronic and mechanical brainchildren of our species subject us, their creators, to selection pressure. "The 11 year olds of today," observes a report from the Carnegie Council on Children, "are the computer terminals of tomorrow."

It is true, of course, that inanimate objects do not dictate ethical norms, but people who deal intimately with such objects can, and do. Also, just as fields of knowledge may be classified by the personalities of the people who create them, so can ethical preferences.* It follows that good and evil for the inhabitants of a technocracy is likely to be defined in terms consistent with the technological-scientific image of man, and that the determination

---

*Already known to Mother Goose:

Soldiers brave, sailors true,
Skilled physicians, Oxford blue,
Crooked lawyer, squire so hale,
Dashing lover, curate pale.

of what is good for man will be judged on equal footing with the determination of what is scientifically true.

Such policies, however, are based on a category mistake, as follows. Science searches for truth, which may be revealed by proper means, for its existence must be assumed to precede its discovery. But unlike the vistas of changeless eternity envisioned by Plato, adopted by science and revealed through experiment and theory, moral choice, to which ethics is to be the guide, derives from man's capacity to determine a yet undetermined future. Hence, moral choice is not an act of discovery but one of creation. Identifying the ontic status of scientific truth with that of ethical stance undermines the idea of responsibility. The flaw in such a view, in the words of Georgescu-Roegen lies in "the refusal to see that in a domain where prediction is impossible it is foolhardy to believe that there are means by which man can achieve some chosen ends and only these."

Striving for the well-being of humanity may certainly be an ethical enterprise, but believing that such well-being may be achieved without appeal to the passionate faculties of man is again a category mistake. Insistence on utopias controlled by science neglects the unresolvable conflicts of the mind, and hence, it leads the disciples back from their mental present to something of a creature present. This regression is exemplified by such traits as the narrowing of temporal interest from historical continuity to a frantic presentness. Advanced technology, and the societal framework in which it flourishes, have turned the oppressive medieval emphasis on tradition and being into an equally oppressive emphasis on no-tradition and becoming.

We found earlier that whenever the experience either of being or of becoming overwhelms the other, the conflict that determines nootemporality lessens, and lower level temporalities become dominant. For individuals, such regressions manifest themselves in ecstasies and may be refreshing. But groups can ill afford to lose their nootemporal capacities collectively, because of their scale. One whirling dervish may be useful: a nation of whirling dervishes may spell the decline of the nation. Yet we

have encountered the advocacy of regression in the teachings of
certain religious ethics, and will find similar regression championed through the awesome power of machine civilization. The
utopian world of Marxism, a state on earth where there will be no
need for further social struggle, corresponds to a fairy tale
temporality where time has no arrow. Writing on entropy, value,
and development Georgescu-Roegen remarks that in the Marxist
state

> as in heaven, man will live forever thereafter without the sin of
> social hatred and struggle. It implies a radical change in man's
> nature, nay, in his biological nature. More precisely, it requires
> that man should by some evolutionary reversal be degraded to the
> status of other animals ....

Georgescu-Roegen argues from a different vantage point
than does the theory of time as conflict; the isomorphism of
argument is the more interesting.

Indeed, when a group of humans functions without the
benefit of nootemporal tension, it runs the danger of becoming
detached from its position in the open-ended hierarchy of unresolvable conflicts. Here Arthur Koestler's Yogi and Commissar
meet in a coincidence of contraries, but their ethical instructions
remain as different as are the characters of Buddha and Faust.
Whereas the Yogi's preference for a timeless present-as-being
tends to produce stagnation, the Commissar's preference for a
timeless present-as-becoming leads to the conditions aptly named
"future shock."

Certain elements of conduct articulated in the West during
the middle decades of the 20th century suggest a revolt against the
major symbol, and probably the major historic cause of "future
shock", the Protestant pragmatic ethics. This revolt may be
associated with the figures of Orpheus, Narcissus, and Dionysus –
sensuousness, beauty, and celebration, rather than with the figure
of Prometheus the Forethinker (who provided man with the skills
of articulate speech, of building of houses, and of making shoes
and bedding). The dimensions of this revolt are limited; its great

importance is not in itself but rather in its function as a rolling stone that starts an avalanche.

The American counterculture, an apt term coined by the sociologist Theodore Roszak, represents the confluence of reactions to certain elements of the Protestant pragmatic ethics, such as cleanliness, punctuality, penuriousness, monogamy, and diligence: all of them useful virtues of the merchant class. Against the ideal of productivity advocated as useful and necessary by St. Augustine, Luther, Calvin, Marx and Marx' many children, the subcultures advocate individuality and creativity as the socially most desirable features of conduct.

The revolution against traditional ethics represented by the American example began in the 1960's with the children of plenty in command of their parents' wealth, of their society's tools, but with no overt obligation to the past. They were following the sound of a new drummer, albeit a rather uncertain and ill-identified one. The long range effect of the counterculture would have been negligible, had its spirit not communicated itself to a much larger group, to wit, to the women's movement of the Western world. Again, the well-to-do, well educated minority of American females, women in command of man's wealth and their society's tools but with no overt obligation to the male, became champions of the ascendency of woman, or more important, of the reinsertion of the feminine among the patterns of praiseworthy conduct.

The unrest of some of the middle-class American females would have also remained almost as inconsequential as that of the counterculture of the young, had it not, in its turn, communicated itself to a more universal sphere, namely, that of a reevaluation of praiseworthy sexual conduct. The ramifications of an emerging new sexual ethics are far reaching, and will be dealt with later.

As the 20th century enters its last quarter, the ethical direction of the industrialized world, from the political point of view, may be figuratively represented as the ghost of Galileo, back to haunt his persecutors. Among the reasons for which he was pursued one issue loomed high: the frightening consequences of removing man, and hence his Church and God, from the center of the universe. No one seems to have been alarmed, as far as I can ascertain, by his extensive use of the geometric concept of time.

Yet consider the powerful sweep of the last one hundred years toward the secularization of ethics in the service of the state. This motion was not at all assisted by the discovery that man is but a speck of dust in a vast universe, which contains things untold. On the contrary, such knowledge engenders awe even in modern man. Rather, the philosophies which are now removing the godhead from the universe, and define good and evil entirely in terms of the needs of the state, claim the power of (Galileo's) mathematized science as their foundations. The arguments of materialistic philosophies rest on the belief that predictability, implicit in exact science, is a universal feature of matter, life, mind, society. The world at large is controlled by scientific law. For this reason it is unnecessary, and hence erroneous to assume the existence of transcendental agencies, such as a divinity. But if this be true, then divinely ordained rules for human conduct make no sense.

But beneath this modern form of Enlightenment which insists on identifying praiseworthy conduct entirely through reason, man still lives by and for ideas and remains, not unlike his paleolithic ancestors, only superficially a reasoning animal. Surely, the cruelty which members of our species are ready to mete out and the suffering they are ready to endure in the name of ideas, may those ideas be ever so irrational, is nothing short of the incredible. It is painfully evident that one must evoke not the absence of moral judgement in man but rather its presence, if one is to account for the frenzy of the ethical animal. Only a creature with unreasonable inner conflicts, and in desperate need of guidance, could be so unfulfilled that he would strike out in all directions that his creative genius might conjure up, even if his actions be clearly deadly for himself and for others.

With established religions on the retreat and political teachings clamoring for recognition, we live in an age when faith is ready to grasp everything and believe anything for the sake of belief, as the mind struggles with its conflicts within the confines of individual selves.

3   The Evolutionary Office of Morals

The possibility that the moral sense of man which underlies
the coherence of society is itself a product of natural selection was
treated extensively by Darwin in *The Descent of Man*. In the
thoroughly evolutionary stance of the theory of time as conflict,
the evolutionary and hierarchical character of morals is to be
taken for granted. Thus, if you were an intelligent queen bee, it
would be your sacred duty to kill your fertile daughters, or else
you would be committing an unpardonable sin. But people are
not bees, and hence no satisfactory meaning can be assigned to
my general claim, as it concerns man, unless this crucial question
is answered first: what precisely is the evolutionary office of
morals?

One answer is suggested by A.O. Wilson in terms of
biophysical dualism.* He sees the sources of man's moral stance
in the nature of life. He searches for the human biogram (those
aspects of behavior by which individuals increase their Darwinian
fitness through the manipulation of society) and assumes that
among its rules the roots of ethics ought to be identifiable.

This sociobiological search begins with a model of ideal
societies, defined as those which lack conflict and possess the
highest degree of altruism and coordination. On such a scale
colonial invertebrates are the most perfect; their individual members (zooids) are completely subordinated to the whole. Insect
societies are next. They function with "impersonal intimacy:"
members recognize castes but not individual nestmates. Their
relative independence goes with a degree of discord. Aggressive-
ness and discord are carried further in vertebrate societies, where
altruism is infrequent and, when it occurs, is usually directed
toward the offspring. Group membership is not mandatory
though often advantageous. Finally, man breaks "the old verte-

*The doctrine which holds that mind and society are biological epiphe-
nomena, but life cannot be totally reduced to physical and chemical lawfulness.
The world, therefore, comprises two major levels: that of the physical and that of
the biological.

brate restraints not by reducing selfishness, but rather by acquiring the intelligence to consult the past and plan the future."

Man also practices reciprocal altruism, or, in more familiar words, the exchange of favors. Since this practice may be extended in time it makes economy possible, whereas trophallactic exchange (exchange of alimentary liquids) as practiced by social insects, does not. As seen by Wilson, human societies approach insect societies in cooperativeness and exceed them in their power of communication, thereby reversing the "downward trend" in social evolution "that prevailed over one billion years ...." This downward trend is a corollary of the division of labor which, in its increasingly advanced forms, leads to competition among different members of a society. The reestablishment on a higher level of the ideal society of colonial invertebrates, according to sociobiology, will be feasible only when "absolute genetic identity makes possible the evolution of altruism."

Let us retrace the same evolutionary path but with a wider perspective, so as to remove the bias of entomology as to what comprises an ideal society.* The claim that absolute genetic identity makes the evolution of altruism possible is isomorphic with our finding about certain features of prototemporal Umwelts. For instance, electrons are completely identical and indeed, any one may be "sacrificed" instead of some other. The great importance of this absolute interchangeability is well known in quantum theory. Electrons, then, would be the most altruistic creatures, were it not for the fact that they have no selves to give up. In an ideal prototemporal Umwelt of absolutely interchangeable people, altruism can be given no meaning for there are no personal identities.

The zooids of colonial invertebrate society may be said to be in a phase of evolution where they have lost their identity, and

---

*We may ask with Niki Jumpei, entomologist protagonist in Kobo Abe's *The Woman in the Dunes*, "Was it permissible to snare, exactly like a mouse or an insect, a man who had his certificate of medical insurance, someone who has paid taxes, who was employed, and whose family records were in order?"

their functions are definable only in terms of their position in the colony. But the colony has not yet totally become a single organism. As in the case of cellular slime mold in the colonial stage, praiseworthy behavior resides in doing whatever the society needs.

For insects and vertebrates, good and evil may also be thought to correspond to pleasure and pain, to reward and punishment. Simple organisms seem to have little difficulty in distinguishing between the opposites, although in higher animals the issue is muddled by elements of guilt and pride. Self-sacrificing behavior in animals can often be shown to have definite value for the species. Hence one might insist that animal altruism is the ancestor of moral stances in man, although for reasons that we shall discuss, altruism cannot be their source. Let us recall what we learned about the policies of nature regarding metastable interfaces. Certain trends identifiable on a lower level open up, or blossom out (both phrases are metaphors) and polarize in many and unpredictable ways on a higher integrative level. In this case, animals display altruism but only man can act with compassion.

It is only with the coemergence of identity, language, advanced symbolism and other nootemporal features, including the human knowledge of time, that the intricate behavioral control known as value judgement, or ethical judgement, could have arisen. The many-sided polarization between good and evil, discussed in the prior sections, is possible only because of the wealth of the internal landscape of the mind. The power of ideals would be difficult to overestimate: too many people have died because of differing ideas. A theory that proposes to identify the sources of moral conduct entirely in terms of its benefit to life will answer such questions as: Why do I defend my children? But vitalistic monism cannot answer such questions as: Why do I defend my "brain children?" The question of good and evil, though surely harking back to certain practices of life, must be examined in terms of its role in the economy of the symbolic continuities that populate the nootemporal Umwelt.

It is true enough that the preservation of life does take

precedence over the preservation of ideas under many conditions, but the limits of the biologically motivated behavior are usually quite clear. Let me elucidate this statement by some examples.

If only one lifeboat is available, it is the women and children who should take it, because even a single surviving male can impregnate several hundred women in the course of a few years, whereas it takes nine months to carry and deliver one child, and several years before that child becomes self-supporting. The behavior of the male in yielding his place in the lifeboat is an example of altruism: life protecting life. But sacrificing the love of wife and child and giving up all earthly goods because of compassion for a man crucified two millenia ago could hardly have its parallel among primates, or even among insects.

Or consider the role of heterosexuality in the affairs of man. A female's genes are represented only in the offspring she herself produces, whereas the male's are represented in the progeny of every female he inseminates. Eggs are more expensive than sperms. There is, therefore, an understandable tendency for females to focus on the survival of the offspring, which is a long-term enterprise, whereas males focus on the successful insemination of females, which is a short-term adventure. The overt, behavioral manifestation of this asymmetry may be found in certain traditional, sex-specific roles. Whereas women, looking for men, tend to seek the whole, males looking for females are easily swayed by the part. But the love poetry of Valéry, Rilke, d'Annunzio, or Eliot or the words of Petrarch, Goethe, or Byron are terribly far removed from the egg-sperm relation. When mapped into the nootemporal domain, the heterosexual process becomes thoroughly modified. In the words of Wilson, since it is non-genetic it can be "decoupled" from the biological systems and considered to be simultaneous with it. Wilson also perceives sexual differentiation as an antisocial force because the antagonism that comes from generic differences can only be contained by mores, but not put into direct use by evolution. It is somehow overlooked that this antagonism is the most powerful single creative force in the affairs of humanity, and that the most important and interesting aspects of good and evil in man belong to the "decoupled" nootemporal system.

Or consider the ritualized ways in which animals guard their status as prey or predator. When the rituals of mutual bondage cross the interface between life and mind, they become dynamic, creative bondages between persecutor and persecuted: Judas and Christ, the Public Prosecutor and Rubashov. On the stage of the internal landscape of the mind, predator and prey can become the same, and a man may distinguish himself from animals by his power to kill himself.

We do not know what the working morals of the caveman were, but they served him well for a few million years. His norms of behavior must have been stable, even if not as stable as the preferred behavior of cockroaches, which have served them for perhaps a billion years. Why, then, were the proven tenets of caveman morality abandoned for the will-o-wisps of historic man, when the latter counsel systematic bloodshed and cruelty, and produce mainly anxiety, with only occasional states of satisfaction? Does this change only improve Darwinian fitness, or, perhaps, does it do something else as well, while still remaining within the functions of evolution by natural selection?

Martin Luther King had a dream of a better society, as have had many people before him. He called for the adjustment of the norms of praiseworthy conduct so that it might lead to the fulfillment of his dream. As have many prophets, thinkers, and poets of all ages and places, he advocated adjustment of a kind that could not possibly be interpreted as adaptive steps. For instead of leading to equilibrium by way of redress, the proposed adjustments were more likely to produce disequilibria in need of yet later redress.

The suggestion comes to mind that the office of the good is not only to protect the status quo (the survival of the child or the idea) but also to produce conflict, possibly of a certain peculiar quality. There is nothing in the millenia of experimentation, sketched earlier under religious and political visions, that would

JOF: I see them, Mia! I see them! Over there against the dark,
stormy sky. They are all there. The smith and Lisa and the knight
and Raval and Jöns and Skat. And Death, the severe master,
invites them to dance. He tells them to hold each other's hands
and then they must tread the dance in a long row. And first goes
the master with his scythe and hourglass, but Skat dangles at the
end with his lyre. They dance away from the dawn and it's a
solemn dance toward the dark lands, while the rain washes their
faces and cleans the salt of the tears from their cheeks.

Ingmar Bergman, *The Seventh Seal*

To dream the impossible dream,
To fight the unbeatable foe,
To bear with unbearable sorrow,
To run where the brave dare not go,

To right the unrightable wrong,
To love pure and chaste from afar,
To try when your arms are too weary,
To reach the unreachable star ...

Joe Darion, "The Quest"
from *Man of La Mancha*

THE WORKING MORALS of the caveman served him well for a few
million years; his norms of behavior must have been almost as stable as
the preferred behavior of cockroaches which has served them for
perhaps a billion years. Why, then, were the stable, proven tenets of
caveman morality abandoned for the will-o-wisps of historic man, when
the latter counsel systematic bloodshed and cruelty, and produce mainly
anxiety, with only occasional states of satisfaction? Does this change only
improve Darwinian fitness, or, perhaps it does something else as well?

There is nothing in the millenia of religious and political experimentation that would point to a body of teaching that guaranteed secure
and lasting balance. The kind of adjustments advocated by prophets,
thinkers, and poets of all ages and places have seldom, if ever, led to
equilibrium by way of redress: they usually produced disequilibria in
need of yet later redress.

The suggestion comes to mind that the office of the good is not
only to protect the status quo (the survival of the child or the idea) but
also, and perhaps mainly, to produce conflict, possibly of a certain
peculiar quality.

Ingmar Bergman: *The Seventh Seal.* Photo, courtesy Janus Films.

point to a body of teaching that guaranteed secure and lasting balance. Even the miraculous homeostasis of Cathay must have been found wanting, for why else would the revolt of modern China have arisen?

Perhaps the evolutionary office of morals is two-fold. On the level of life it protects life; on the level where it is detached from life it creates disequilibrium and conflict. Let us, therefore, distinguish two types of moral obligations, and reconcile the idea of good and evil with the hierarchical structure of nature. Let us class a man's protecting the life of another as *duty*. A man has duty as a father to his children, as a husband to his wife. And let us class a man's protecting his "brain children" as *responsibility*. All higher obligations, such as taught by political philosophies, advanced religions, and laws that assume free will, are responsibilities. A dog does adore his master and there are clear-cut selective advantages to his behavior: its faithfulness is that of duty. But it is in the domain of responsibilities, that is, of higher obligations, that the story of humankind truly unfolds.

It is a reasonable assumption that since Homo sapiens appeared on earth, it is not his brain or body that has changed most but rather his skill to use the capacities of his brain. It is not equally reasonable to assume, yet I do not hesitate to do so, that there has also been a progress in the mental capacities of the best of mankind. A further suggestion then comes to mind: the office of moral behavior has been the creation and maintenance of disequilibria, or conflict, that tends to lead to the emergence of newer forms in the Umwelt of the mind, rather than to the collapse and simplification of those forms.

There are tremendously great difficulties in identifying specific instructions that can be used to create conflicts resolvable by emergence. The spectrum of relevant adaptive features in man is so wide that it amounts to an innate moral pluralism; hence biology can be of no aid. The difficulties are endemic because ethical stances, as I argued earlier, do not constitute discoveries of preexisting truths, but rather creative determinations of yet undetermined futures.

Thus, though hints may be had, no technical prescription can exist that would remove from man the responsibility for his conduct, and guide him with certainty toward the constructive resolutions of his conflicts. The existential stress of the mind will be resolved by different people at different times and places into different rubrics of good and evil. But for all men, between the extremes of God and Devil, there stretches a continuous range of concerns. The human moral predicament is a spectrum of cares whose resolution into good and evil remain relative to individual, time, and place.

In spite of this relativity of ethics, and due to the confluence of many and complex causes, there is a distinct possibility that the family of man will come to form a single coherent unit. This would be a world-wide organism, repeating on a higher level the experience of the colonial invertebrates, or perhaps of insect societies. Such an organism would function on the basis of "impersonal intimacy," recognizing only distinct castes but not distinct individuals. From the point of view of this organism, our arguments on responsibility would have to appear both useless and dangerous.

4   The Individual and Society – Part II

William Shakespeare had his own way of expressing the unresolvable conflicts of the mind, those generated by the difference between the expected and the encountered. He said

that the will is
infinite, and the execution confined; that the
desire is boundless and the act slave to limit.

This conflict may be resolved by collapse, it may be maintained, or it may be solved through the emergence of a new integrative level.

In *Romeo and Juliet* society is punished for its assumption of preeminence over the individual, whereas in the plays of

Beckett the individuals are punished for assuming preeminence over society. In our epoch tragedy, the classic image of the individual fighting against the powers of society or tradition, or just against blind fate, is quite unpopular. Its replacement is something of an applied pathos (as in "applied geometry") which, in its purpose and effects, resembles uncomfortably closely the chemical "passion surrogate" of the *Brave New World*. It is as though the power of the individual had run out of steam.

The social dynamics of our epoch are volatile, the structures kaleidoscopic; one can sense a universal malaise attendant to the ferment of minds and bodies. The conditions of the noetic world of our age resemble those of life's cul-de-sac which, perhaps a million years ago, brough about the emergence of the mind. At that time it was life that could not adapt sufficiently rapidly to the changed environment which its own evolutionary success had created. This time it is the individual mind which cannot adapt rapidly enough to the changed environment created by the collective actions of many minds. It is as though individual selfhood had reached a dead-end, for it is unable to maintain its unresolvable conflicts while remaining within the confines of the nootemporal Umwelt. The conditions which thus prevail resemble those states of nature that were identified earlier as metastable interfaces. From the numerous examples that could be marshaled in support of this belief we shall examine a representative few.

○ Earlier we traced the origins of an emerging, new *sexual ethics*. The new code of praiseworthy conduct as regards intimacy with members of the opposite sex is the product of many forces, among which I wish to mention the need for, and the possibility of, birth control.

The ease with which procreational and recreational sex can be separated is in the process of undermining those ethical codes whose justifications, overtly or covertly, are biological. In an increasingly larger area of the world individuals may fall in love and fall in sex quite independently. We have found earlier that the origin of heterosexuality is perhaps the major driving force in the search for knowledge. It follows that the direct modulation of sexual behavior constitutes a more profound interference in daily life than can be accomplished through the control of bread or political freedom. The instincts for the perpetuation of the self and of the species may now be separately directed. Social control of procreation need not inhibit sexual satisfaction, hence, libidinal energies that previously could be directed against authority, or else sublimated into creative action, need not ever be mobilized.

○ Simultaneously, with this subtle but powerful reorientation of sexual energies, we find a change in women's status which might best be described as *Eve unbound.* In the third world and in the Communist countries, with their energies directed toward a reordering of economic structures, women's lot is at least as heavy as it traditionally has been: an inhuman mixture of physical and mental anguish, finished only by early death. But in the socially advanced West, as well as in industrialized Japan to a much lesser degree, there is a drive for minimizing sex-coded roles. This is the direction in which the world seems to be going. The change is fired not by compassion, but, from the point of view of society, by the fact that the equalization of women permits increased versatility in the division of labor, that is, in the production of specialists. There are several corollaries of this trend.

One is the surfacing of an increased number of tough and cruel females, in the historical tradition of the cold and intelligent Cleopatra, the power-hungry Roman Theodora, and the blood thirsty empress Tzu-Hsi. The new tyranny of the feminine is in the ascendancy. Another corollary is the decline in the estimate of the masculine, with the attendant confusion of the young male and the increase of adolescent sexuality in adulthood. Yet another corollary is the decrease in the estimate of motherhood and its replacement with "bottlehood," as masses of males and females march toward the status of Aldous Huxley's freemartins.* The Ewig-Weibliche of Parcelsus and the Great Mother of Greece,

*In the *Brave New World* infants are not born but decanted from bottles.

Bottle of mine, it's you I've always wanted!
Bottle of mine, why was I ever decanted?

Rome, and the Orient are on their way to the closet of skeletons. Monogamy, having its roots in the necessities of securing and raising offspring, is about to lose its raison d'être as experimentation with simultaneous multiple sexual relations increases. The new ethics wherein Eros and Agape are not permitted to maintain their creative struggle points to the eventual replacement of all meaningful political freedom by sexual liberty.

○ We found earlier that technology and science are examples of communal enterprises that extend the bodily and mental functions of individuals, permitting results that no one person, working alone, could possibly achieve. Societies built on science and technology demand a sophisticated *division of labor* with interchangeability of specialists in specific skills. We reencounter here the conditions which are kept in such high esteem in sociobiology: the perfection represented by insect societies. A smoothly run technocracy also functions on the basis of "impersonal intimacy" wherein members recognize occupations but not individuals within occupations. In the most advanced of human societies, as among colonial invertebrates, one would expect a drive toward completely subordinating the individual members to the whole as a unit.*

Human anthills are thus in the making: some drab, some grey, some in technicolor, but anthills just the same. Kafka's Castle is being exported from its natural habitat to the New and even Newer Worlds, distributed and promoted by the unholy alliance of opportunistic ideologies, international terrorism, and multinational corporations. As in the gastrulation of embryos, so different groups of man are shifting around in search for, or anticipating their final position in, the family of man.

The egalitarian ideas of the modern West have their sources in Christian soteriology, intended for the salvation of the individual. These ideas of equality of conditions are slowly changing to

*Cf. the Enlightenment view of hidden design in nature. "The division of labor from which so many advantages are derived is not originally the effect of human wisdom .... It is the necessary, though very slow and gradual consequence of certain propensity in human nature." Adam Smith, *The Wealth of Nations,*

demands for guaranteed equality of results, a reasonable corollary of interchangeability of people. But being a member of a human anthill forces the mind to operate in a prototemporal Umwelt of indistinguishable elements. In such an environment the cherished character of human time is nowhere to be found. The emptiness left by the loss of identity can be filled only partially by increased sexual freedom. The ancient remedy for the loneliness of the self has been that of the intimacy of a few bodies. With the interchangeability of people and the separation of recreational from procreational sex the trend is toward many bodies.

Stafford Beer once estimated that genetic learning is eight billion times slower than noetic learning. If the present population of the earth were to double, genetic changes could be produced at a rate comparable to noetic learning. Then humanity would have a true opportunity to become a single organism. But whereas biological evolution does not transmit acquired characteristics, social evolution does. So the spirit, somehow, would still run ahead, though no one knows where.

It has been said that in the Heavenly City all shall be one, but I doubt that the hope of that prophecy corresponds to the promise of the modern state.

○ Earlier we spoke of "the horror visited upon man's mind if circumstances suggest loss of identity." I have just reasoned that the societal environment of interchangeable people is that of prototemporality which, as we know, is characterized by statistical laws. Likewise, in a technocracy of specialists, social care tends to become statistical and hence lose its human quality. In the words of the Massachusetts poet John Boyle O'Reilly, one has

> Organized charity, scrimped and iced,
> In the name of a cautious, statistical Christ.

This degrading, downward step communicates itself to the whole spectrum of problems that pertain to *suffering and death.*

Violence, for instance, is probably not more frequent in the experience of our contemporaries than it was in the lives of our ancestors. However, through the magic of communication media, it has become simplified and standardized: through statistics it becomes acceptable.

Suffering is a uniquely human capacity which includes pain, but goes much beyond it. Since suffering signals a danger to the integrity of the self, it has an inherently personal significance. As Ivan Illich has eloquently stressed, there exist classical virtues among the individual's responses to suffering: patience, courage, forbearance. It is through suffering as a first hand experience that man learns to have compassion for others. The numerical social security of the modern state transforms the compassion for suffering into an enterprise of pain killing, with loss of the dignity of the individual. One may ask: are we to stop the pain-reducing efforts of medicine? The only humane answer is: no. Medicine has performed near miracles for individuals and masses of people. But this is precisely the point. The family of man has entered a state of transience, wherein our actions, informed of garden-variety good will and employing the best technical regimen, nevertheless, and inevitably, rob the individual of his dignity.

The poor countries of the world represent about two billion people, almost two-thirds of the world's population, with millions of humans who exist in the shadow of death by starvation and disease. The 200,000 new mouths to be fed each day exemplify the fiasco of life's attempt to resolve its unresolvable conflicts by biological means alone. The methods used by Norwegian mountain villagers to avoid the dangers of starvation are suitable for groups of 20 or 200 people. When dealing with two billion people, personal meaning has no place; the solution must be statistical. The madness of the wildly reproducing biomass thus leads to an inevitable reduction of human society to the level of insect societies.*

---

*One might counter that it is appropriate for institutions to consider the individual as a statistic, as long as individuals within the institution consider one another as individuals. The point which is being made, however, is that the impersonality necessary for social progress unavoidably leads the individual to considering himself and others as statistics.

There is an outstanding and ironic example here of the new uses for old processes across the biotemporal/nootemporal interface. Anti-birth control policies have been championed by the Church of Rome, intended surely to increase the number of good Catholics around the globe. But this policy of protecting life promotes the demise of individuality and the rise of a world where secular religions must surely replace transcendental religions, including Roman Catholicism.

"Great religions," writes Ivan Illich in his incisive criticism of the secular city "provide a social reinforcement of resignation to misfortune, because they offer a rationale and a style for dignified suffering." Whereas traditional religions were appropriate to the noetic world of individuals and spoke the language of individuals, the new religions (Communism, finance and mercantile capitalism) are dedicated to the sanctity of technical progress and must speak in languages appropriate to the societal integrative level. Thus, for instance, the concern with aging in America is being replaced by geriatrics, the science of treating the old. Concern with love and intimacy are being replaced by the study of interpersonal relationships and of sexual methodology. The result, again, is a shift from the person as an individual to a citizen who fits a statistical category.

The awful distance between the dying and the living is a corollary of having been a person, and a hallmark of life's irreversible journey. That distance is being increased through the ritual death-dance by the physician and his magic machines. The suffering and dignity of death is coming to be subsumed under the phrase "dying policy," according to the *Encyclopedia of Bioethics.* No death is beautiful, all deaths are ugly, but, so the living suppose, a shared room and sky with those one loves may perhaps imply, to the dying, the power of continuity. In contrast, the death of a man after long suffering, with tubes and needles in him, in an oxygen tent, is the death of a man exiled by an insane government to die alone on the surface of a distant star. Not that being surrounded by people is sufficient: dying of hunger in Calcutta, of exhaustion in the mines of South Africa, or in the labor camps of Soviet Russia, do not constitute desirable alternatives.

To death there is no alternative, but if humanity enters a
new level of complexity some people, at least, will have to die
twice: once as individuals, and once as bodies. This is exactly what
happened to *Doctor Zhivago*'s Lara.

> One day Larisa Feodorovna went out and did not come back. She
> must have been arrested in the street at that time. She vanished
> without a trace and probably died somewhere, forgotten, as a
> nameless number on a list that afterwards got mislaid, in one of
> the innumerable mixed or women's concentration camps in the
> north.

○ We are witnessing a *media revolution*. The camera is rapidly
replacing the pen, noise is rapidly replacing silence, and books are
being pushed out by voice and visual means of communication.
But I do not judge these changes to be regressions to pre-literacy.
They suggest, rather, that picture and voice are more appropriate
means for the societal Umwelt than is the written word. Free and
compulsory education that teaches people how to read and write
does not, and did not, prevent this shift. Passing judgment (as
distinct from assimilating what is being read) is hard labor,
practiced by few. The camera, that at one time was reputed to be
unable to lie, has made the objectification of tendentious fantasy
absurdly easy and believable. Furthermore, on all levels of
communication, we observe an encroachment by number, and a
corollary insistence on the quantification of experience.

There is a frightening increase of environmental noise in all
well-inhabited regions, much above what is unavoidable. It comes
from prostituted word and music, and from machine-based joys. It
is my experience that higher environmental din tends to inhibit
thought and I cannot avoid the feeling that the appreciation of
such din stems from the desire to inhibit individuation. It is
perhaps not by chance that noise, loss of identity, and use of
number, arose together as some of the hallmarks of the sociotemporal Umwelt: communication by number is the most efficient
language for a noisy environment.

○ Earlier we imagined two archetypal characters playing on

the stage of selfhood: the Observer and the Agent. They were analytic projections of a single set of functions, that of the self. On the stage of nationhood these imaginary actors separate into two traditional personalities: the Prophet and the Statesman. They need not be, but usually are, two distinct people. Their interactions, dialogues, enmities, trusts, and mistrusts are good indices of the dynamics of a society. The trend in our metastable epoch is toward *silencing the Prophet.*

It will be recalled that the Prophet (the Observer) speaks with the "inner voices" of man. His words are often those of our paleologic; his utterances are often charismatic, addressed to a world seen as essentially predictable. The utterances of the Statesman (the Agent) pertain to what appears possible in a world where the future is unpredictable.

A society without a continuous struggle between its Prophets and Statesmen is likely to lose its creative dissonance just as the individual self vanishes if the Agent and the Observer coalesce. If the Prophet (or the Statesman) is silenced, a class of societal functions disappears. This class comprises processes that are isomorphic to those of the unconscious, between body and mind. I believe that in the absence of the creative dissonance between Prophet and Statesman the stability of nationhood is threatened. Together with other forces already considered, the silencing of the Prophet intensifies the metastability of our epoch, and increases thereby the chances that man will have to mold into a human anthill, or else relapse into tribal forms of life.

The silence of the Prophet is a reality of daily life for one-third of the people on earth. Enormous, inert populations remain profoundly indifferent to such occasional dissidents as may arise. To the ears of these masses anything but expected cliches sound irritating. There has emerged, in our century, a pattern of behavior which may be associated with the idea of the totalitarian mind: preference for social passivity, an underdeveloped sense of individual worth, the absence of responsibility, in general, a flight from the freedom and burden of selfhood.

Because of hunger, illness, and poverty, moving under the weight of history, and counseled by ethnic temperament, modern totalitarian systems operate in a silent conspiracy between the regimes and the masses; we can observe a symbiosis between prey and predator. Since the daily routine of life does not provide legitimate outlets for expressing complaint, or rage, or even spontaneous admiration, the citizens carry repressed emotions, to be unleashed at the pleasure of the state. Certain inevitable reactions to the loss of individuation, such as the widely spread alcoholism in the Soviet Union, reinforce social passivity and guilt, and make political manipulation of the masses easier.

The path to the creation of the totalitarian way of life need not be paved, however, by overt repression. An inert population of indistinguishable citizens, without social responsibility, may also be created through conditions and practices which are not dictatorial in their original intent. For instance, the Prophet may be silenced by being replaced by the futurologist or its ilk (equipped with computers which, so the story runs, make him more efficient than the computerless Prophet). Also, the dignity of man may be degraded, slowly but surely, along any of the avenues which we have been discussing. This is the direction in which the free world seems to be going.

The rapprochement among the powers of the earth is not one of avowed governmental forms, but one of totalitarian mentality. Corresponding forms of government are likely to come later, following, as they usually do, the temperament of the people. According to an old addage, nations always have the kind of kings they deserve. The final state of the first, second, third and nth world may well be alike, though reached by different paths.

○ During the last few centuries much igenuity has been expended in the socially advanced countries of the world on divising a means whereby individual and societal interests may be reconciled. Foremost representatives of these efforts are the American Constitution and the *United States of America* itself.

The Constitution was drafted in a curious mixture of experimental spirit, hope, selfishness, and compromise. Thus, Benjamin Franklin remarked at the Constitutional Convention that "we are, I think, in the right road for improvement, for we are making experiments." Thomas Jefferson hoped to keep America an agricultural nation. He feared that people and government would remain virtuous only "as long as there shall be vacant lands in any part of America. When they get piled upon one another in large cities, as in Europe, they will become corrupt as in Europe" and another system might be necessary. "I have sworn upon the altar of God," he wrote in 1800, speaking both as a Prophet and a Statesman, "eternal hostility against every form of tyranny over the mind of man." "The tree of liberty," he added seven years later, "must be refreshed from time to time with the blood of patriots and tyrants. It is its natural manure."

For a number of reasons, foremost among them being the built-in capacity of the Constitution to provide for its own reinterpretation, the governmental structure based on that document has been able to maintain a reconciliation between individual and collective interests. John Adams described this operation, referring to the balance among segments of government, as that of checks and balances. By chance, historical good luck, and the Yankee genius for compromise, the government so constituted permitted and encouraged a permanent revolution, and has succeeded, more or less, in containing the conflict between the interests of the citizens and those of special interest groups.

But Americans are now "piled upon one another in large cities" under conditions to which the Jeffersonian ideals cannot apply. The sonnet of Emma Lazarus, engraved on the Statue of Liberty in 1886, says, in part,

> Give me your tired, your poor
> Your huddled masses yearning to breathe free,
> The wretched refuse of your teeming shore,
> Send these, the homeless, the tempest-tossed to me:
> I lift my lamp beside the golden door.

In ninety years this message has deteriorated, with unpleasant frequency, into such abominations as the following "Invitation to New York Corporations Thinking of Moving Their Headquarters to Fairfield County:"

Give me your energetic, your rich;
Your privileged yearning to breathe free,
The executives of your teaming shore,
Send these, the achievers to us:
We lift our lamp beside the golden door.

With no apologies to Emma Lazarus.*

Checks and balances as a principle of government is being challenged by doctrines that define the individual entirely in terms of its role in the societal collective. The attack, as it unfolds daily, comes from right and left, from the combined powers of radical egalitarianism and conservative capitalism. In a striking way, the centrifugal and centripetal excesses that challenge the stability of the American system reflect world-wide conditions. The underlying cause is a fact of industrialization: the exosomatic evolution has turned production into an intrinsically communal enterprise.

The only thing that could change this trend would be a return to the use of endosomatic instruments, such as was called for by Alexander Solzhenitsyn in his *Letter to the Soviet Leaders.* This is a personal manifesto of tremendous power, written in the spirit of the Russian messianic tradition. But the world at large hears a very different voice. Reducing the scale of life to that of the individual, changing the economy mainly to manual production, abandoning the internal combustion engine (even temporarily), and favoring the intimacy of village life (on traditional Russian patterns) are not the ideals that move the nations and the governments of our age.

Developing and developed countries alike are under the spell of a mystical trust in modern science and technology. It is conceivable that the inversion of evolutionary roles among matter, life, mind, and society (see below) will force upon mankind a renewal economy, and its corollary, a no-growth society. But because of the reasons already given and yet to be discussed, this

*Reprinted by permission, *Fairfield County Magazine,* Edward J. Brennan, Publisher.

would only make mankind poorer and life more restricted. It would not be likely to change the trend toward the decreasing significance of the individual. A return to artisanry, to the family, to slow travel, and to the silences and noises of nature is certainly an imaginable development of mankind, but only if the unstable conditions of our age were to lead to increased stability by means of collapse rather than emergence.

In America and around the world there is an increasing polarization between demands for uniformity and cooperation, and demands for tribal, cultural, religious, ideological, professional, ethnic, linguistic, economic, racial, sexual, national – and all other imaginable separate powers. The situation is one of metastability between an increasingly aggressive pluralism and an increasingly aggressive singularism.

○ Society may and does suppress certain memories, and there are no reasons to believe that this has not always been the case. The most frequent form of supression (if one dare make any generalizations whatsoever) involves reinterpretations of the past. For instance, many pre-Christian celebrations became Christian feasts, and the appearance of many a leader, Muhammed being one, were seen by his followers as having happened just at the right time. It remained to our epoch to prefer a *world without history* or, more precisely, the claim that the past is irrelevant.

Daily life in the industrializing countries is increasingly informed of a pervasive impermanence of all things; the waning interest in durable possessions is a paradigm of the situation. The corollary narrowing of interest to the present is not an invention of the "now generation;" the same is implicit in the "business present" (both Western and Eastern style) which pays only lip service to past and future, as it demands results right now.

It is interesting to recall that the French Republican Calendar, promulgated in 1793, intended to recast the rhythm of society. This was to be accomplished by modifying the dating system, the length of the week, of the month, the subdivision of the day, etc. so as to make these correspond to the values of the French revolution. But France returned to the Gregorian calendar in less than thirteen years. Unlike the precipitous attempt of the French Academicians, the industrial reevaluation of temporality works slowly, but deeply, as part and parcel of that change from the individual to society which we have been tracing. The modern calendar is being adapted to the demands of the modern state. The symbolic significance of its divisions is replaced by the practical concerns of a cyclic existence. The past of this existence is hardly more than a tenuous drama, its future is subject mostly to science fiction stories. The "Toy of the Month Club" might have taken Aldous Huxley's instructions as its motto: "Feeling lurks in that interval of time between desire and its consummation. Shorten that interval, break down all unnecessary barriers." With the noetic conflict lessened, man regresses into a history-less existence.

○ There has been a substantial and rapid increase of *mind-made surrogates* for inanimate and living things, and even for feelings. Artificial flowers and flower-scents are pushing out flowers and their scents; mind-made milk (chemicals that look and taste like milk), and orange juice made from oil products and water, are beginning to replace milk and orange juice. The environmental degradation is not a matter of the increased biomass alone, but of the radical increase of chemicals that do not occur naturally. Pesticides and additives, paints and preservatives, propellants and spray cans add their share of ruin to that caused by the effluents of industry and exhaust of engines. Machine music, informed of systems rather than feelings, is considered "in," as is man-made music that sounds as though it were machine-made. Erotic equipment may be recognized in the sculptures of Babylon and Egypt, and was quite popular as "godemiches consolateurs" in 18th century France. But only we can boast radio-controlled dancing eggs, placed inside the vagina, to be used "at work, on the beach, or on the bus home from school." And mind-made-minds are advertised as improved replacements for the one made by organic evolution. I sense the fulfillment of Blake's *London* of 1794

In every Infant's cry of fear,
In every voice, in every ban,
The mind-forg'd manacles I hear.

Consider now that matter cannot change rapidly enough to keep up with the needs of life. Throughout organic evolution the surface of the earth was only lightly populated, so new forms of the increasing biomass could expand into existing ecological niches. But as living matter further increased in volume and in power, an inversion of roles became significant: the earth had to adapt to life, (if not universally, then locally), rather than vice versa. This inversion was characterized by recurring crises in food supply. With the appearance of man came demands made upon the environment by small groups of humans. Eventually there came the demands of the collectives of minds upon their environments, which include both matter and life. But life cannot evolve rapidly enough to adapt to the imaginary country of the mind, hence we have the problems of the human body not able to live in a mind-made environment. This is one of the crises of modern society. Finally, the scientific-industrial society of our epoch is now changing so rapidly that the individual minds cannot adapt to it as their new environment. This is the crisis of the emergence of the societal integrative level.

This hierarchy of inversion in evolutionary roles amounts to a generalized Malthusian Principle: life outruns matter; man outruns life; mind outruns man; society outruns mind. And just as the ordinary Malthusian conditions influence, even though they do not determine, the structure of human ethics, so the nature of the good as held by society is influenced by the inversion of these evolutionary roles. The tremendous increase of mind-made surrogates to satisfy the needs of life, man, feeling, and even thought, is the societal means of redressing the inversion of evolutionary roles.

○ I have already pointed to the difficulties implicit in the *uniqueness of a global society:* it cannot define itself against other global societies. It cannot check its autogenic imagery against its peers, friends, and foes, and thus try to determine which plans are possible, which ones probable, and which impossible. Many instances of the politico-economic conditions of our epoch do not even have a history: there can exist no guidelines based on experience according to which the problems might be solved. Change in life styles is rapid. Not surprisingly, there is much trial and error experimentation, often imbued with a spirit of recklessness. This experimentation combines a spirit of nihilism, a curious mixture of radical disbelief in orderly change, with a faith in the magic powers of technology, and, in a large part of the earth, with a disclaim of individual responsibility. But any trial and error test, if miscarried, may be fatal for the world community, though not necessarily for the survival of man. The frightening swings of hunger and wealth, the increasingly richer promises of plenty contrasted with equal promises of horrors and holocausts, all suggest that the interface is indeed metastable and could easily be set into explosive oscillations.

It is reassuring that at least the periodic regression from the phenotype to the genotype is still with us. We learn from Frederick Winsor's *The Space Child's Mother Goose* that

> Foundations shake,
> Computers break
> And Science goes be-bop,
> But baby's joy
> Is still the toy
> With foolish ears that flop.

In Part I of this section I argued that the regularities of the sociotemporal Umwelt, if they are to be intelligible, must be expressed in a language appropriate to that integrative level. I am not imagining that citizens of that new world would be more perceptive than their ancestors were, or even better informed; the opposite is more likely to be the case. What I am saying is that nootemporality might appear to the new communal forms of discourse as an inadequate mode of coping with the world, insufficiently developed for an understanding of some of the

(now) puzzling aspects of human time. In the language appropriate for the societal Umwelt, some issues that pertain to death
control, to free will and destiny, to the role of the individual versus
society, to issues which now elude even the keenest of minds, may
be so stated as to become answerable. But whether or not there
will be a call for answers, is another issue. It is conceivable that
speculative curiosity may be an expandable part of societal
language, which shall deal only with economical and production
instructions. The population problem might become manifest in a
language of imperatives without reflective content.

In the hierarchical structure of the world contemplated by
the theory of time as conflict, epistemology itself is hierarchical.
Working within the confines of nootemporal language it is not
possible to delineate which portions of our problems fall into the
unanswerable category. We cannot tell which of our questions
may be answerable "from above," and which shall, or at least
may, find satisfactory explication within the world of the individual mind.

Let us then do the best we can, which is the taking of a
Socratic view of the question of society and the good.

"Now is the time that we are going," said Socrates, "I to die
and you to live, but which of us has the happier prospect is
unknown to anyone but God." The Socratic puzzle has not been
answered by the secular city but only bypassed; its unresolvable
conflicts are on their way to partial resolution through the
interchangeability of individuals. The theory of time as conflict
can say nothing about possible detailed development along this
path, or even whether there shall be development. It can only
assert that the present global status of man is metastable, in the
sense discussed.

The metastability of the noetic/societal interface is not to be
identified with any of the classical ideas of revolution. It is not the
revolt of the hungry against the well fed, or that of one race
against another or one nation against another, and not even one
ideology against the other, although it includes all of these. Rather
it is a revolt against the unresolvable conflicts of the mind. These

can lead either to the emergence of a new integrative level, or to a collapse to the level of individual minds grouped in tribes and families, or even to regression to the biotemporal, leaving the earth for the beasts that crawl, fly, and swim.

Hence, the march is not to the tune of Dies Irae, or to the Marseillaise, not even to that of the Communist Internationale, but to something much more elemental: the march is to the voices of the caged mind. We are in a metamorphosis wherein, as Kazantzakis put it "life has grown more savage, and the gods grown more powerful."

If all this sounds ominous, and it does, one might take the advice of Justice Oliver Wendell Holmes, a man profoundly dedicated to permanent revolution as a societal way of existence. "Have faith," he wrote, "and pursue the unknown end."

XI The Freedom of the Beautiful

THE HUMAN NEED to admire and create beautiful objects, events, and utterances is as ancient as the mind, as ubiquitous as language, and as intimate as selfhood. But there is no such thing as a uniformity of taste. On the contrary, the spectrum of what has been judged beautiful at different epochs and places is immensely wide. The non-uniformity of aesthetic judgment is much more impressive than are specific rules that might have guided peoples' tastes.

It is the thesis of this chapter that there do not, for there cannot, exist universal and stable instructions as to what constitutes consummate beauty, but there is nonetheless an underlying need for it. Among the three principles that can guide man's behavior, the true is the most restrictive category, the good is less so, while the beautiful pertains to free experimentation with yet undetermined forms, events, and utterances. This freedom is made possible by the capacity of conscious experience to rove unrestricted among, what I shall call, the moods of temporality, and then return, as it were, and tell us about its journey.

1   Timelessness

Earlier we spoke of the Augustinian uncertainty principle according to which the mature experience of nootemporality obtains from the balance of two opposing traits: time felt and time understood. The unresolvability of this opposition is manifest in the existential tension of the mind. This tension is that of individuation, of the self, of a burden from which, as far as one can tell, man has tried to escape as soon as he was able to experience it. One class of such escapes is associated with the ordinary idea of timelessness.

It is an empirical fact that as one leaves the middle region of the Augustinian balance for altered states of consciousness, a feeling of selflessness, elation, timelessness may be experienced.

For instance, attention may be restricted to an unchanging single item or process. The being-like component of the existential stress is thus emphasized and the becoming-like repressed; the unresolvable conflict of individuation lessens and the Umwelt of the mind becomes one that resembles the eotemporal. The direction of time vanishes; the definition of nowness and selfhood loosen up. This state of elation may be called the *ecstasy of the forest.* It is also possible to restrict one's attention to ceaseless change, as through the *ecstasy of the dance.* This is another very ancient method of lessening the burden of individuation. The tension of selfhood lessens, the direction of time retreats from consciousness, and the Umwelt of the mind is again eotemporal. Natural selection has taken advantage of these ecstasies and incorporated them into its methods that lure humans into preserving their species. Sexual intercourse offers in one fell swoop the timelessness of the dance, followed by that of the forest, in the *ecstasy of the bower.* The eotemporal Umwelt that often characterizes the state of mind after intercourse easily blends into that decreased level of self-awareness which is the hallmark of sleep.

Religious and political systems have employed many methods that help decrease the unresolvable stresses of selfhood. The promise of lessening tension attracts and keeps people in the fold. The *ecstasy of the chalice* of Christianity is one means whereby the individual may experience an oceanic feeling of eternity and then journey back to self awareness and serve the ideology of the chalice.

Alcohol was the not-so-secret weapon of the slave traders of the 17th and 18th centuries, and has remained that of some of the slave states of the 20th century. Heroin has been a weapon of Oriental militarists of our own epoch, both Japanese and South East Asian. Drug-induced loss of the sense of time has often been regarded as divine in origin. In our secular epoch drug-induced states of consciousness are often described as transcendental. Let me class the plant-produced conditions (whether by chewing, smoking, or fermenting and drinking) as the *ecstasies of the mushroom,* in honor of fly agaric, poisonous and sacred, which keep the elves who live underneath them dry even in the worst of downpours. The *ecstasy of the mob* places the individual in a prototemporal situation because, from the point of view of the collective, he is countable, but otherwise indistinguishable from other members of the mob. Appropriately, the frenzy of mob action appears to the participant as an experience of a long present without well-defined futurity or pastness, wherein the gnawing issues of hope, memory, guilt, and responsibility have little or no significance.

In these and many other widely available and traditionally practiced ecstasies neither is the feeling of timelessness one of the total absence of time (such as would accompany a loss of consciousness) nor should these states of mind be confused with conditions of atemporal chaos (suct as might possibly be the case in the painful confusion of certain schizophrenic experiences). Rather, they represent regressions from the nootemporal Umwelt of the mind to Umwelts of lower temporalities, to wit, to the biotemporality of animals or the eotemporality of certain creatures and things.

Scanning the ecstasies just listed might suggest that regressions to timelessness have been used mostly to favor man-devouring Molochs; and they have been. But for each type of regression to lower temporalities one can also point to examples that illustrate how ecstasies have been employed in reaching for "the light, the light, the seeking, the searching, in chaos, in chaos." The ecstasies of the forest, the dance, the bower inform and generate knowledge and through it, create and destroy civilizations. The ecstasies of the chalice, the mushroom, and the mob make and break great individuals and social systems. The power of being able to wander among temporalities does not foreshadow darkness of doom, nor does it guarantee the light of salvation. The results of time-roving depend, instead, on the controls that regulate and channel the power of ecstasies.

I believe that the admiration and creation of beautiful objects, events, and utterances is the most universal and powerful single means through which the free play of the mind may be organized and related to the other enterprises of our species. The soma regresses to the gene periodically so as the bring forth new and different bodies; our conscious experience regresses periodically into states of dream so as to bring forth newly integrated world views; likewise, we must revert to the unbridled play of the beautiful so as to create new reality.

2   The Moods of Time

The hallmarks of different Umwelts, when thought of as features of worlds in which we might be required to live, bring forth different constellations of feelings. These, the affective dimensions of the temporal Umwelts, I shall describe as their *moods*. Moods tend to be as distinct as are integrative levels and, while they last, stable. Emotionally we may be in one mood, while intellectually we can still grasp many things.

In this section I shall try to illustrate these moods by examples from the graphic arts and from music. Then I shall argue that the beautiful, the ugly, the frightening, etc., are composite moods constructed from the elementary moods of the temporal Umwelts.

(a)   The Atemporal and the Prototemporal Moods

Atemporality, as I have reasoned, is not isomorphic with nothingness or non-existence but rather with the idea of an empty set, or of chaos. Likewise, the most primitive, and hence, probably the most frightening of moods, is that of the atemporal. It is a mood of chaos, of emptiness. I do not believe it to resemble any of the ecstasies mentioned, because those described are but muted senses of time. The best I can do is to quote the Japanese poet Nishida Kitaro who seems to speak about the emptiness of the atemporal mood when he writes that

> The bottom of my soul has such depth
> Neither joy nor the waves of sorrow can reach it.

Above this emptiness we find the prototemporal Umwelt with its countable but not orderable elements. Things are never definitely here or there; events are never definitely now or then. Those familiar with contemporary music may think of the second movement of Alban Berg's *Lyric Suite:* its construction is unpredictable, the sounds are purposely incoherent; the composer called this movement "Desolato." Or consider Dadaism: a revolt not against any specific method of organization, but against organizability itself. Yet Dada seldom succeeded in reaching pure prototemporality, because a striving for meaninglessness remained meaningful, just as Berg could not get rid of all formal structuring.

However, consider the paintings of Jackson Pollock or Mark Tobey (e. g. Pollock's *Phosphorescence* or Tobey's *Universal Field*). They represent prototemporal inarticulation, emerging from the primordial sources of feelings. The mood of aleatory painting relates to the mood of mature selves as the structure of vapor relates to the organization of the brain. In Kurt Vonnegut's *The Sirens of Titan* we read about an army of millions parading on an immense steel plate. The soldiers are in uniform, move in unison as the snare drum declares:

> Rented a tent, a tent, a tent;
> Rented a tent, a tent.
> Rented a tent!
> Rented a tent!
> Rented a, rented a tent!

Prototemporal moods seem to communicate with the archaic levels of our mind. They correspond the the world from which aggregate matter, life, and man himself evolved. For the mind, a descent from its noetic environment to a prototemporal one is a feat of anguish. Functioning in a prototemporal environment, whether demanded by art or by politics, involves the dissolution of the self, that is, an attack on personal integrity.

When the genius of the artist suggests a desolate, prototemporal background out of which organization is about to arise,

(corresponding, perhaps, to the prototemporal/eotemporal interface), the impression or mood can certainly be joyous; at least, so I
reacted to Tobey's *Voyagers III* and to Vasili Kandinsky's *No. 58*
and *Light Picture No. 188.*

(b)   The Eotemporal Mood

Observe any of the multitude of early Egyptian reliefs
where each portion of the body is represented in its most telling
perspective. The shoulders are seen from the front, as are the eyes;
but the face and the hands are in profile. Usually one says that the
artist used different perspectives; he asked his model to turn or he
imagined the motion. Instead of this interpretation, clearly biased
in favor of the spatial, we can give one favoring time. The artist
collected in a single view what he had seen at different times. He
found nothing inappropriate about combining views obtained at
different times into one, undifferentiated present, as though his
attention were able to expand to several hours, or days. The
prevailing mood of such reliefs is eotemporal.
    Jumping 40 centuries to the Renaissance and a few thou-
sand miles West, to Florence, we may think of Jacopo da Pontor-
mo's large canvas *Story of Joseph.* The various phases of the life of
Joseph are placed in a coherent spatial setting as though they were
simultaneous. Jump another 200 years to 1754 and over to
England, where William Hogarth made a delightful engraving
called *False Perspective.* It is a country scene with a brook, a
bridge, an inn, a church, animals, people. As a line of sheep walk
into the distance they get larger; there is a nearby cuckoo sitting
on a distant tree, and a traveler on a faraway hill gets his pipe
lighted by a candle held out from the window of a nearby inn.
Nothing is inconsistent in this picture if we permit the artist to
timesect the world, to combine into one view many vistas he has
seen at various times. This world is one of pure succession where
futurity and pastness combine. Picasso's celebrated *Les Demoi-
selles d'Avignon* (1907) is clearly eotemporal in its dominant
mood; it resembles early Egyptian reliefs.
    The subject of a painting need not be a close likeness of a

natural object for that painting to be eotemporal in its dominant mood. Marcel Duchamp's *Coffee Grinder,* Georges Braque's *Piano and Mandola* and Fernand Leger's *Smokers* are abstract representations in the cubist style. Cubist doctrine demands the combination on the canvas of various views which belong to the same structure, but are not perceivable at a single instant; cubism insists on an eotemporal mood.

The eotemporal mood permits a larger variation of reactions than does the prototemporal mood. Perhaps the eotemporal is the mood of the early man, of the child, and of the child in the adult; it is surely the mood of fairy tales that happened "once upon a time," and of the Bible: "In illo tempore."

In the prototemporal mood it is easy to claim isomorphism between spatial and temporal representations. Those familiar with the Berg music and the Pollock painting that I mentioned will immediately appreciate their kinship. The isomorphism is a reminder that on the lower integrative levels spatiality and temporality are only weakly differentiated. It is possible but more difficult to find such analogies with eotemporal moods, and they are quite impossible for the higher moods. No painter could reproduce the pictures of Musorgski's *Pictures at an Exhibition* by listening to the music, and no uninformed composer could produce Musorgski's music by examining the paintings.

(c)   The Biotemporal Mood

The biotemporal world is more complex than the physical; the biotemporal mood is more elusive than the primitive moods.

In Kurt Vonnegut's *Slaughterhouse Five* we anchor ourselves in time with ordinary future, past, and present, but only for a while. Soon we are yanked into another situation for another while, then into yet another. In each segment of the story references are made to past and future, in ways that make it impossible to decide whether what we are reading is the description of a present happening, whether it is memory, or perhaps prophecy. Vonnegut's characters are presumably in full command of their faculties, yet they act as though they were organisms simpler than man. The total mood is that of nowness emerging, in fits and starts, from a matrix of pure succession. The result is disturbing.

The same emergence may be looked upon differently. One may try to show the excitement of creation. This is the mood of many works by the Catalan painter Joan Miro. His *Beautiful Bird, Two Personages, Dog Barking,* or *Harlequins* suggest to me the emergence of life, the immense journey from inorganic matter to life.

Approaching the upper regions of the biotemporal world, we encounter emerging selfhood. Salvador Dali's *Apparition of Face and Fruitdish on Beach* leaves me with the feeling that, were I to wait long enough, the chalice-and-face that occupies the center of the picture would change into a chalice or into a face but would refuse to remain both, unlike the well-known reversible images which alternate in brief periods of time. Dali's surrealism shows the dawn of identity. The fathers of surrealism held that their art depicts a world which is in some ways superior to that of our noetic world. This, I think, is upside down. A more appropriate name for surrealism would be archaic realism, for what makes surrealism interesting and powerful is its appeal to the early evolutionary Umwelts of our minds and feelings.

A literary description of emergent identities was given by Marcel Proust, master of the twilight zones between self and non-self, in his *Remembrance of Things Past.*

> ... in my own bed, my sleep was so heavy as completely to relax my consciousness; for then I lost all sense of the place in which I had gone to sleep, and when I awoke at midnight, not knowing where I was, I could not be sure at first who I was, I had only the most rudimentary sense of existence, such as may lurk and flicker in the depths of an animal's consciousness; but then ... memory ... would come like a rope let down from heaven to draw me out of the abyss of not-being ... in a flash I would traverse and surmount centuries of civilization, and out of half-visualized succession of oil lamps, followed by shirts with turned-down collars, would put together by degrees the component parts of my ego.
>
> Perhaps the immobility of things that surround us is forced upon them by our conviction that they are themselves, and not anything else, and by the immobility of our conception of them.

(d)   The Nootemporal Mood

The temporal moods considered thus far comprise the affects, cognition, and conation of the nootemporal self with respect to conditions that are associated with different temporalities. But we have no plateau from which the features of a nootemporal mood could be beheld, by investigating the affects of this (higher) consciousness with respect to conditions that are associated with nootemporality. This, the highest of moods, can only be put together by degrees, as Proust put together his ego. I believe that there is a way of doing so, provided we are willing to settle for nothing more than an approximate but useful scheme of aesthetic experience.

It is well known that each time we examine an object our gaze follows a spatial scanning pattern. Psychological studies of artistic vision suggest that another scanning process, however, is also involved. This one is a mental scanning. Before an articulated, final image emerges into consciousness, our visual perception runs through undifferentiated stages of dreamlike structures, and experiments with different memory and perception systems on different developmental levels. According to Anton Ehrenzweig, each act of visual perception tends to recapitulate the ontogenetic development in the visual motor perception of the child. All of these stages can be identified in the perceptive experiences of individuals, and some of them have also been identified in the development of perceptive modes in the history of art.

But it then follows that our consciousness must continuously oscillate among temporalities that correspond to the developmental levels of perceptive moods, all the way from a dream-like sense of unity to the keen recognition of distinct identities. But whereas the scanning-gaze pattern can be easily tested by photographing the small motions of the eye, the many moods one would expect to be associated with the temporal scanning of objects are privileged disclosures. We know of no measurable physiological equivalents to the vast spectrum of human feelings. Furthermore, to the extent that the temporal scanning may be partly or totally unconscious, the viewer or listener himself might be quite unaware of the component parts of his final judgment. To muster support for the time-scanning hypothesis, one must proceed indirectly.

The reasons which I find strongly suggestive of the validity of our ceaseless "time travel" may be found in the analysis of a peculiar kind of feeling, known as the uncanny. To make my point, it will first be necessary to detour to a discussion of the uncanny.

Freud has shown that the sources of the feeling of the eerie (uncanny) reside in reencountering conditions which were once familiar, but then became repressed and forgotten. When such forgotten conditions are reencountered they appear to be both familiar and unfamiliar; hence they are surrounded by an aura of incomprehension and mystery.

Consider, for instance small children with their bears, dolls, parents, and siblings. In a practical way, a small child can certainly distinguish between her mother and her doll, but since her Umwelt is not yet sufficiently differentiated, her relationship to objects and to people have much in common. The loss of a toy bear may be as tragic as the loss of a person, because in the child's world bears and people share many fundamental attributes. It is not an exaggeration to say that the child considers them in many ways identical. Later the child learns to repress this feeling of identity and she banishes from her mind the theory that dolls are people; this theory is never to be mentioned again seriously, as outmoded scientific theories are banished by communal repression. When much later the child, now fully grown, encounters a very life-like robot or a robot-like person, or sees something that might be either, she is likely to experience the feeling of uncanny.

Tampering with matters that were once familiar but later repressed is usually discouraged by forceful taboos. We are forbidden to handle people as though they were dolls; we come to regard bodily discharges which were once considered as parts of ourselves, as not parts of ourselves; and it is socially unacceptable to copulate with a formerly familiar corpse. There are equally strong taboos against tampering with temporal categories. Once the hierarchy of identities, causations, and temporalities becomes integrated, conscious experience comes to be protected, so as to assure that what was repressed remains repressed. The taboos shield the great discovery of the mind: that of temporal passage. Whenever one of the taboos is violated, one encounters conditions formerly known but later repressed, and a feeling of the uncanny is likely to be experienced.

For instance, it is not permissible for grown persons to believe in almighty kings who know all the past and all the future; hence, events which suggest correctness of a prophecy appear uncanny. Neither are we permitted to believe that frogs can change into princes, or statues into live women; hence, connections that smack of magic causation tend to appear eerie. The actions of our parents were once incomprehensible and overwhelming; as we grew, our parents shrank and became life-sized. Whenever we encounter the alleged results of vast and incomprehensible powers, we experience a feeling of the uncanny. We may take any of the features of primitive Umwelts that were once known to the child, and were believed by him to be real, present them on a canvas or in a book as though true, and we have an eerie painting or story. Let us consider a few examples.

The famous *Last Judgment* of Hieronymous Bosch, from the turn of the 15th century, shows a world where the familiar rules that govern life and behavior are grossly violated. A creature on two huge feet with two dead fish for bosoms and a metal box for a head is something the mind might conjure up, because it is the mind's task to generate images. But if ever dreamt of or hallucinated, the image would be repressed as unreal. Yet the world of such creatures remains faintly familiar; we have met them in our nightmares. There is another character on the canvas: it has clawed feet like a bird, but is a person, and is frying another person in a frying pan. This is rather an unceremonial auto-da-fé, one which must have looked uncomfortably close to reality to those who were pursued by the Inquisition. Such images also lurk in the depths of every soul. Hence, the *Last Judgment* is eerie. The same may be said about *St. Anthony Tormented by Demons*,

Martin Schongauer's etching, also from the 15th century. Distort-
ed organisms poke the Saint in ways that seem to communicate
with the untamed animal evident in every healthy child, and not
so evident but definitely present in every adult. Demons themselves are creatures of the imagination, which, when checked
against modern reality, are found not to have been produced by
organic evolution. That is good, because they are awfully uncanny.

Giuseppe Arcimbolo's *Fire* (mid 16th century) shows things
that belong in the inanimate world (candles, oil lamps, cannons,
burning logs) combined so as to make up the profile of a man.
Upon first encounter, the picture is the impression of a weathered-
looking male. As our perception scans the various levels of its
Gestalt, our consciousness grasps that the man is only an appearance, an epiphenomenon of fiery things. Suddenly, the painting
becomes uncanny. Although the mechanical metaphor of man
and life was quite popular in the 16th and 17th centuries, and
little boys sometimes see people as though put together of
mechanical parts, the idea that I, and my friends, and all my loves
are robots is vehemently rejected by the mind. Arcimbolo's *Water*
is again the likeness of a man made up of creatures without mind:
fish, lobsters, crabs. It suggests a blasphemy, it violates a taboo,
and it is certainly uncanny. In Max Ernst's *A Spanish Physician*,
the hierarchy of Umwelts has gone mad: his creatures are neither
animals nor people but both; they ought not to exist. Casper
David Friedrich's *Cross and Cathedral* confuses, with marvelous
skill, the relative positions of the earthly and the divine: the
building is just too abstract and regular to be earthly, yet too worn
to be divine. Like a doll that is too life-like or a woman who
moves like a machine, the Umwelt of the picture is between two
established categories; its (presumed) reality violates a taboo.

Finally, I want to mention Picasso's celebrated *Guernica*. In
that painting the identity of each and every organism is challenged; and identity is a cherished privilege of man. We developed it as a species and learned it as children. Identity came to
each of us through hard work, as I amply stressed in earlier
chapters. *Guernica* attacks the integrity of man and beast; all

continuities are torn asunder. There is a clear regression from identity to namelessness, from integrity to dust, from life to death. The painting is an image of the collapse of the nootemporal and the biotemporal Umwelts into the physical world; the pain and suffering that accompany this violation of the autonomy of mind and life is overwhelming.

Any number of objects which I could have chosen, from the fine arts or from literature, do strike us as uncanny. There are elements in them which were once familiar, then repressed. But if we cannot actually recall, or even say exactly why something is uncanny, yet feel strongly that it is, this is evidence that our attention continuously scans the various moods of the objects. What we can bring up into consciousness is the total impact of sight or hearing.

The reason I have selected to discuss the uncanny is that there happens to be a body of understanding about it. If we indeed "travel" among the moods of art objects and come up with a total impression of the uncanny, does it not stand to reason that what we judge as ugly, frightening, soothing, neutral, or beautiful is also put together from impressions gained while traveling among the temporal moods? I suggest that what we declare to be our aesthetic experience comprises the totality of all lower order moods, weighted by personal preference and cultural filtering.

3   **The Aesthetic Adventure**

The beautiful designs of many butterflies have very mundane purposes. Unlike their homelier relatives, brightly colored butterflies often contain substances distasteful to predators. The easily noticeable colors educate individual predators; they eat only one pretty butterfly, and learn not to touch the rest. Many butterflies have also developed wing patterns for private communication: The patterns are visible only in the ultraviolet, hence to other butterflies but not to predators. In the life of the butterfly, aesthetic and utilitarian judgments cannot be distinguished.

I would not know what butterfly emotions may be, but there is enough evidence to suggest that more advanced organisms possess a wide spectrum of feelings. Yet, except in a very few examples in birds, any significant appreciation of beauty apart from usefulness does not seem to exist. Animals could not share the biotemporal or eotemporal moods associated with sense experience, because these moods are the reactions of the mature mind to those respective Umwelts. Due to the absence of an imaginative mind, all but perhaps some anthropoid apes lack the ability to alter their feelings. Humans can do so for a number of reasons, such as for a change, or on account of criticism. The difference lies in the capacity to produce symbolic transformations of experience. Perhaps some animals do experience their own animal-temporal-moods, but I see no reasons why they should be driven to experiment with them, to travel among them, and much less to return and give a (symbolic) account of their adventures. Animal drawings suggest that while they do make systems and may enjoy the physical activity of drawing, they do not care for the results and do not perceive the representation of feelings in objects.

For the ability to draw systematically, the hand must be guided by inhibitory controls that involve temporal delay – memory, expectation, and the mental present. Between the visual or auditory stimuli and the corresponding manual or other bodily reactions, a pause must be inserted so as to permit the modulation of the final action by feeling and reason. Only through the slow evolutionary development of pictorial and vocal representations of the human intellect and emotion, did that external, linguistic system come about, whose internal, parallel system we recognize as artistic quality.*

*The idea that art is creative both in its execution and in its appreciation, and hence is necessarily a human enterprise, has been eloquently defended by the philosopher, biologist, and educator, Nathaniel Lawrence. The notion of evolution, he writes, "has remained too long captive to a biological ideology. The inescapable result is to treat of humanity under the major category of its body. The educational outcome has been to suppose that you have done enough for the creative needs of the student when you have sent him to school and provided him with paper, crayon, and clay. Here is as sure a set of villainous ideas as I know of, a gang of wreckers of human self-awareness. I say, throw the rascals out."

Studies in the developmental psychology of art have revealed that by the age of 7 or 8 years the child can become a participant in the artistic process without the need of further substantial qualitative reorganization of his mental and emotional development. Beyond this age he can, of course, acquire increased skill, grow in his experience as a human being, and become acquainted with artistic tradition. When beholding the moods of objects of art, of certain events, or of conditions, the onlooker or listener may readily identify with those moods, because all people have been shaped by essentially similar processes of unconscious identification throughout uncounted generations. Scientific gifts can also be evident at an early age, but this is rather rare. The earliest substantial insight recorded by Einstein goes back to his middle adolescent years, whereas poetry, music, and paintings of great sensitivity are frequent in children no more than ten years old. Also, for art the past is only a guide, whereas scientific knowledge is cumulative and the student must first learn what has been done before him. The road to our knowledge of the lower Umwelts, best represented by the hard sciences, is longer than that to the higher Umwelts, because with the inanimate world we have no feelings to share.

Until the birth of scientific psychology, dreams were the concern of art rather than of science. We might turn this around. The aesthetic adventure of Homo may be seen as related to his capacity to lift from his dream-like experiences those images and sounds which, in combination, bring forth the compound mood that his intellect wishes to represent. Artists of all ages have been suggesting images with which all people may structure their daydreams. Memory, which is a type of daydream, and the manifest content of dream images have much in common. Consider that in many dreams, though events do not happen all at once, there exists nevertheless an awareness of past and future that makes the mood eotemporal, in the sense discussed earlier. ("I knew that he would be around the corner, and he was.") We found earlier that the same mood characterizes memory images.

Let us return to the butterfly in whose life the functions of beauty and usefulness cannot be distinguished. And let us recall from chapter five that certain single functions which have two

aspects on the biotemporal integrative level open up into two
distinct functions when projected across the biotemporal/nootem-
poral interface. In this example, aesthetics splits off from pragmat-
ics, and man separates the beautiful from the useful. The common
roots are, nevertheless, maintained: nothing useful is totally
without beauty, and nothing beautiful is totally useless.

It is the aesthetic gifts of man that are first to meet all new
worlds, and it is through the community of moods that these new
worlds first reach other minds. From mathematics and the hard
sciences to ditch digging, warfare, architecture, and music, it is the
ethically and aesthetically satisfying (though not necessarily the
good and the beautiful) to which we must first appeal for guidance among the images created by the mind. Perhaps we should
refine and enlarge Louis Pasteur's aphorism: chance favors the
prepared mind and the sensitive soul.

4   Music do I hear ...

or is it poetry or plain prose? Their origins are common in
the utterances of hominids and they have a long and distinguished
ancestry.

Chimpanzee troops sometimes go on "carnivals:" they
shout as loudly as they can and drum with their paws on tree
trunks. The purpose seems to be to keep the dispersed animals
together. Gorillas display chestbeating and thumping, often with
predictable sequence and rhythms. The function of the display is
advertisement, threat, or just play. Individual whales sometimes
have their own variations of a species-specific song that can last
up to 30 minutes; the purpose of the songs is probably that of
mutual recognition as the whales migrate in the Atlantic Ocean.
There is evidence to suggest that improvisation in the song of
certain birds is analogous to human musical composition. The
skeleton of the song is innate, a population specific overlay is
acquired by learning, and individual variations are invented by
the bird. Cranes, sea eagles, quail, grebes often sing in duet. The
purpose of the bird songs ranges from territorial danger and

sexual calls, to pure improvisation for no apparent purpose other than enjoyment. But in spite of the importance and vast range of animal sounds, and of the occasional signs of inventiveness, the separation of music-like and speech-like sounds in the animal kingdom is either not possible or is, at best, very marginal.

I would like to speculate that speech and music became differentiated only when, and as, the evolving mind learned to separate predominantly emotive utterances from predominantly intellectual ones. In due course the two groups of sound came to define two cosmologies. Earlier we dealt with the cosmology that is made possible by articulate language. It was concluded that the intellect, individually and collectively, defines the noetic universe by complementarity: society and the cosmos define each other. Our interest here is in the cosmology made possible by emotions: not independent of the intellect, but definitely distinct from it.

A piece of music and its elements, tone, rhythm, melody, all consist of sound frequency modulations. A tone comprises oscillations from a few cycles per second to perhaps 16 KHz. Rhythms are frequencies compatible with the motion of the human body, such as in dance; melodic sentences encompass minutes; a musical piece itself is a repeatable unit that might last a few hours. Thus, the frequency range of the "ars bene modulandi" takes up the important middle region in the cyclic order of life. These frequencies all run simultaneously, both in predictable and in unpredictable modes. Within the resulting rich sound we may identify the hallmarks of different temporal moods, even though no single mood may be meaningfully extracted from music as a whole.

Rhythmic repetitions determine par excellence an eotemporal world of pure succession; the unpredictable or only probabilistically predictable elements determine prototemporality. The spectrum of sound brings into play all the physiological and psychological faculties that we normally employ in time perception and in constructing our sense of time. In music we are called upon to use very short term and long term memories as well as expectations. Unlike the visual arts that modify, behold, and create an external reality that appears to be independent of the viewer, music and poetry enter directly into the audio loop that helps define the self. The tension and relaxation of musical metaphors and their multilevel play are paradigms of existential stress. For this reason, music can reflect emotive cosmologies, complementary to the cognitive cosmologies of science.

The affective content of some compositions may be identified with the timeless ecstasies discussed earlier. Such identifications are conditioned by cultural filtering and personality, but this does not lessen the validity of the general claim that music of all epochs has spoken to man about timelessness, albeit in different dialects. I wish to list here some examples, first studied and discussed by L. E. Rowell.

O *Zangetsu* (Morning Moon) by Kinto Minezaki suggests the ecstasy of the forest. Its melody, for the traditional Japanese ensemble of koto, shamisen, shakuhachi and female voice, is infused with the atmosphere of the timelessness of the country dawn. In the context of being-a-time, one might also think of the

THE SPECTRUM OF MUSICAL SOUND brings into play all the physiological and psychological faculties that we normally employ in time perception and in constructing our sense of time. The tensions and relaxations of the musical metaphors and their multilevel play are paradigms of existential stress. Musical sounds enter directly into the audio loop that helps define the self.

> For the good are always the merry,
> Save by an evil chance,
> And the merry love the fiddle,
> And the merry love to dance.
>
>   W. B. Yeats
>   "The Fiddler of Dooney"

Hokusai, Katsushika (1760–1849). Edo period. *The Lion Dance* Courtesy of the Smithsonian Institution, Freer Gallery of Art, Washington, D.C.

orchestral introduction to Benjamin Britten's opera *Peter Grimes*. It is a metaphor of the quiet contemplation of the infinity of the sea, stretching out far enough to merge with the sky.

○ Maurice Ravel's *Bolero* is a brilliant orchestral tour-de-force, a seamless unit of insistent, hypnotic crescendo over a single repeated rhythmic pattern. It is surely a musical image of the ecstasy of the dance.

○ The orchestral introduction to Richard Strauss' opera *Der Rosenkavalier* is a vivid description of the ecstasy of the bower. The careful listener will detect both male and female orgasms musically transformed, and will note a delightful post-intercourse languor upon which the curtain rises to find the Marschallin and young Octavian – in bed.

○ The ecstasy of the chalice may be exemplified by the waves of Gregorian chant. Its melodies speak of a period when eternity was the most important category of time. This writer once heard "Ubi caritas et amor, Deus ibi est," a hymn for the washing of the feet on Holy Thursday, sung by the monks of the Abbey St. Pierre de Solesmes, and he did feel one with the God of Christ. The pagan version of this ecstasy might be the orchestral prelude to *Das Rheingold* of Richard Wagner. The music is that of the creation of the universe, a Teutonic cosmogony: the Rhine river welling up from its primordial source.

○ *Ork Alarm* is a mind-blowing, experimental rock composition by the Turkish composer Jannik Top. It is a musical image of the ecstasy of the mushroom; it might even be that of Turkish poppy.

○ Finally, what can better illustrate the ecstasy of the mob than the stirring music of *La Marseillaise:* "Allons, enfants de la patrie! Le jour de gloire est arrivé!" From the point of view of "la patrie," "les enfants" are countable but otherwise indistinguishable members of a prototemporal ensemble. In terms of "la jour de gloire" there is only a long present, without future or past. The Umwelt is eotemporal. The total mood is that of timelessness.

Long before music or poetry came about, there was dance. Only a few creatures are poets and musicians, but all living things dance.* I find this fact so impressive that I wish to define poetry as dance put into words, and music as the twin sister of dance. My use of metaphor puts the definition closer to poetry than to prose but, in an epistemological aside about art, I wish to argue that metaphors are the earliest and simplest forms of explanations.

To the child, the moon may be a funny face that sometimes looks through the window, sometimes hides behind a cloud. To say that it is a sphere just like the earth is nonsense. The earth is a flat place with houses next to our house, and with grandfather's place in Kansas which, as everybody knows, is flat.

According to ancient tradition the reproductive powers of the male reside in his bone marrow, and so it was even for Shakespeare. But today we know better: reproductive powers reside in the genes. This is our theory; the bone marrow is a fantasy. When Yeats writes that

> He that sings a lasting song
> Thinks in a marrow bone,

we say that this is a metaphor. That the double helix is the material of inheritance might well be judged a metaphor two centuries hence. It is in the sense here implied that the metaphors of dance, music, poetry, and prose are our first guides to reality; they, together, form a cosmology of feeling.

If it is true that every act of visual perception involves the scanning of several Umwelts and experiencing of several temporal moods, even though mostly unconsciously, then we should expect an analogous situation in auditory perception. If there are any level-specific temporal moods to be experienced in a practically infinite number of juxtapositions, then music, poetry, and dance should be able to speaí to us about them. It is up to the genius of the artist and to the gift of the listener or participant to create and

*To appreciate this, we must sometimes see them in fast-time or slow-time motion pictures.

share composite moods brought back, as it were, from journeys among the temporalities.*

I believe that just as language is necessary for the noetic self, so poetry, music, and dance are necessary for the definition of the emotive self. And as the intellectual image of the self defined the noetic cosmos, the emotive self defines the emotive universe. These universes, while distinct, are not independent, "for a tear," observed Blake in *Jerusalem,* "is an intellectual thing." Only a creature with man's unresolvable conflicts would have reason to descend and explore the lower temporalities, create metaphors about his experiences and, with the assistance of his intellect, ritualize them in controlled ecstasies.

5   Tragedy do I sense ...

or is it comedy? ** As do music and poetry, joy and sorrow also have their origins in animal behavior. A dog or a horse, if abused, displays corresponding signs of subjective misery; in animal groups such expressions are contagious and those close by seem to share the mood. Likewise, many animals can express joy, and some can even play tricks on others, including people, then sit back and grimace; this condition can also be contagious. However, I know of no animal behavior that suggests a belief that the chicken's fate is a projection of some universal injustice intrinsic in the world, or else a reflection of the humor implicit in the animal predicament. Only the mind can generalize suffering and joy and see in them certain patterns that infuse the whole of Creation. It remained for man to create the Play. "Sometimes

---

*Why much of contemporary Western music and poetry prefers moods that are heavily prototemporal and eotemporal, rather than the wealth of mixed moods, is but one of the signs of the metastable nature of our epoch, in flight from the unresolvable conflicts of selfhood. Cf.: "The arrow of time, dominant for almost three centuries, is missing in the music typical of this century." (Schuldt).

** I am thinking primarily of quality, and not of form.

everything seems like a long strange dream," says Eva in Ingmar Bergman's *Shame.* "It's not my dream, it's someone else's, that I'm forced to take part in. Nothing is properly real. It's all made up. What do you think will happen when the person who has dreamed us wakes up and is ashamed of his dream?"

Both comedy and tragedy, as understood in the West, originated in Greek drama. Though both imply and utilize the complete spectrum of human emotions and intellect, they have been divergent in their attitudes towards the unresolvable conflicts of man. Whereas comedy as a genre celebrates the successful immediate resolution of problems (by fortunate circumstances, often related to sex) tragedy acknowledges the long range forces which make the conflicts of the human predicament unresolvable. In both tragedy and comedy the spoken voice enters directly and intimately into the auditory loop of self-definition; the dance of characters appeals to the visual faculties of the spectator, and in both types of drama emotions are supplied, so to speak, ready to be used.

But it is the privilege of tragedy to place itself above transient and judicious solutions of difficult situations and to seek (and display) the spectrum of man's unresolvable conflicts. Therefore, it is to tragedy that we shall turn.

It is only an opinion, but one I strongly hold, that tragedy in the realm of speech is akin to music in the realm of sound, because each, in its own way, plays upon the moods of all temporalities. Since the time of Aristotle, and at least partly because of him, tragedy has remained an inductive view of the world, wherein characters are secondary and the plot primary. The tragic experience is a generalization to be made jointly by writer, actor, and viewer (or reader) based on a paradigmatic example. The hero tests and reconfirms our precarious position in the noetic Umwelt vis-a-vis the societal, the biotemporal, and the physical Umwelts. The demands of these integrative levels are contradictory more often than not. The hero continuously compares the expected with the encountered, his needs and hopes with the possibilities; he continuously scans his memories and tries, as well as he can, to control the future through the present. Whereas a single instant of a life can easily look comical, horrifying, or sad, it can seldom be tragic. Tragedy, for its full development, must draw upon the complete hierarchy of temporalitites; it is an art form that demands the fullness of time. In it we are likely to encounter the unpredictable or probabilistically predictable events that characterize the prototemporal, the reversible conditions of the eotemporal mood ("as day follows day"), and a presentness of extreme importance, embedded in the nootemporal stresses of futurity and pastness: "tomorrow, and tomorrow, and tomorrow." Only a creature deeply concerned with, and disturbed by the finity and infinity of time would cry out in response to a fatal challenge to his identity, that "time, that takes survey of all the world, must have a stop." The extreme wealth of the nowness in tragedy is so important that it should bear a distinct name: the tragic present.

Twenty-three centuries have passed since Aristotle died and almost four since Hotspur first died on the Shakespearean stage. The classic insight into man's life as an acting-out of the tragic present is as valid today as it was when first invented. But the modern self has undergone certain changes which have made tragedy, as a form of art, unpopular and well-nigh taboo in the great technological civilizations.

We witnessed a radical change in the description of the universe, from that of a cosmic unit which included man, to that of formulas in astral geometry that include only the lower Um-welts. Since man and the universe mutually define each other, today no less than yesterday, the change in cosmologies corre-sponds to a change in the self-description of man. This path may be traced with reference to the narrowing of the wealth of a tragic present to the whimpering of a creature present. The change amounts to a refusal to roam the disturbing lower temporalities and return from them to the nootemporal.

In *The Physicists* by the Swiss playwright Dürrenmatt, the potential tragedy of free will turns into the immediate easy solution of a comedy. The superb plays of Beckett are cosmologies of people who have deteriorated into organisms of the biotemporal Umwelt, with only vague recall of their humanity. In their worlds, in *Waiting for Godot* or *Endgame,* there are no alternative solutions to the no-solution which informs their existence, and there are no apparent efforts, or conceivable ways whereby these characters would "take up arms against a sea of troubles, and by opposing, end them." The mood of the plays is not nootemporal. The characters suffer like beasts or grasp a passing, happy second like dogs catching tidbits; there is no meaningful self-search, no true identity, no futurity or pastness, no human presentness – let alone tragic presentness. The plays are not comedies – for comedies still command the potentialities of the mature mind. They are anti-tragedies and as such, appropriate for our metastable epoch.

The reasons that could elucidate the demise of tragedy in favor of anti-tragedy or, much worse, of "applied pathos," are rooted in the changing position of the individual in society. Forces stronger than man are identified exclusively with the laws of nature and these, in their turn, are principles neither sympathetic nor antagonistic to man.* The moods of the lower Umwelts are inappropriate for life and mind. Since the idea of a reality, higher than that of man, is taboo, the potentiality of conflict between individual and universal principles vanishes. With the fear of the unknown and the admittance of the unknowable repressed, time has become a commodity. But commodities are like chickens: kept, bought, and sold, they are not heroically offered or withdrawn. In this view of time there are no available means to change individual bravery to inspiring heroism or to acknowledge greatness. In the industrialized countries, the concentration on the business present insists on achievement "now", thereby replacing the time of mature consciousness with that of the child.

The wealth of temporality that informs the tragic present has fallen on difficult days; the taste appropriate for the metastable, transitional state of man of our epoch impugns the pain and the glory of an undetermined future and relevant past.

---

*Art, wrote Joseph Conrad "is not the clear logic of a triumphant conclusion; it is not the unveiling of one of those heartless secrets which are called the Laws of Nature. It is not less great, but only more difficult."

Consequently, its present is too narrow to accommodate tragedy. The unresolvable conflicts of the human predicament go unacknowledged.

6   The Journey of Galatea

To Socrates art was the mirror of the world, an accurate mimesis of what there is. This tradition of creating true likenesses has been inherited by our epoch and is being championed, not by the arts, but by the sciences, where it survives in the idea of mathematical models. Platonic thought, with its demands for precision and for earthly projections of heavenly truths, is still preferred by those who are uncomfortable with that absence of rules which characterizes the exploration of creative imagery through art. According to Plato

> poetry and in general the mimetic art produces a product that is far removed from truth in the accomplishment of its task, and associates with the part in us that is most remote from intelligence, and is its companion and friend for no sound and true purpose.

This is from the *Republic*. The type of knowledge which in our days searches for final truth through intelligence is the work of the scientist, building the utopian republic of the scientific-industrial complex. Plato, champion of the beautiful but not of art, would have been pleased with the symmetries and intellectual reflections of our quantified knowledge of nature.

In a rather non-Platonic way, the poet Frederick Turner sees the function of art and its origins in the labor of the artist who cuts a blaze at the edge of a fixed (known) Parmenedian universe and by so doing, renders it into a Heracletian one for the purpose and duration of his work. Using a different metaphor, the artist, viewer, and listener, by doing what is aesthetically satisfying, chart a lane that did not exist before it was charted. From its primitive manifestations in the protoaesthetics of animals, art speaks of the unpredictable and the new; hence, we might still use the Socratic formulation and see art as a mirror of the world, if we also append a Churchillian exclamation: What mirror ... what world!

If history is any guide, then the moods of artistic creation reflect as well as form the major modes of feeling that characterize epochs and places. What does this say about our own metastable age, caught between the demands of the individual and those of society?

We have already noted that a substantial portion of modern music has proto- and eotemporal moods. We have also learned of the demise of the tragic present in dramatic form and its replacement with something that resembles the creature present of animals. Let us add now two opinions on Picasso's *Guernica*. The first one is by John Fowles, novelist of unusual depth. He speaks of modern art in general, and *Guernica* in particular, through the voice of an old painter who sees our epoch from a distance.

> Turning away from nature and reality has atrociously distorted the relationship between painter and audience; now one painted for intellects and theories ... jettisoning of the human body and its natural physical perception was a vicious spiral, a vortex, a drain to nothingness .... One sheltered behind notions of staying 'open' to contemporary currents .... Op art and pop art, conceptualism, photorealism ... rootlessness orbiting in frozen outer space ... in the bottomless night.

This is an excellent critique of the proto- and eotemporal moods, or even the biotemporal mood, when they predominate in a piece of art rather than form part and parcel of a richer whole.

In Virgil Gheorghiu's novel *The Twenty Fifth Hour* two prisoners of war are being transported in an open truck, packed so tightly that they cannot move, can hardly breathe, and for days must relieve themselves standing.

> "Picasso painted your portrait, just as you are now in this truck, old man."
> "My portrait?" asked Johann. "I can't hear. My ears are blocked."
> "Your portrait," said Traian. "The likeness is as good as a photograph. Seven men occupying the same area in space at the

same time. One has five legs, another has three heads, but no lungs. You have a voice but no mouth, while I have nothing but a head, a head floating in space over a truck ... He paints as if he were taking photographs. Nothing but real life."

The leading trend of contemporary art reflects the general regression of our societal attitudes and individual lives from the nootemporal to lower moods. Perhaps, as earlier arguments suggest, this is one symptom of the metastability of our epoch, a withdrawal of the leopard before it is ready to leap.

Let us now turn explicitly to the idea of the beautiful. In many forms and in many languages this term, or whatever corresponds to it, has been used to describe certain composite moods, or constellations of feelings. These moods are engendered by events, situations, objects – by practically anything and everything that may happen to a person. I would like to tie the idea of the beautiful to the capacity of the mind to wander among the moods of temporal Umwelts, by referring to the figures of two women. One comes from the tradition of Jehovah thundering above the clouds, the other from the tradition of Zeus, doing likewise.

The first woman is known only by the name of her husband, Lot. As Lot, his wife, and his servants took to the hills to get away from the havoc of Sodom and Gomorrha, Lot's wife violated the Laws of her Creator and was changed into a pillar of salt. We may imagine her changing from a woman and mother of feeling and mind to a mindless living body, then to a dead body, and finally to an amorphous rock. Here was a journey of collapse backward along the evolutionary Umwelts from the nootemporal to the prototemporal.

The second story begins with Pygmalion, son of Belus. He fashioned a milk-white image of Aphrodite out of rock and laid it in his bed. Aphrodite, the goddess of beauty, was so moved by this that she brought the statue to life as Galatea who, in due course, bore a son and a daughter to Pygmalion. Galatea's journey was that of emergence among the evolutionary Umwelts from the prototemporal to the nootemporal.

Both of these stories are a bit uncanny, for they imply the operation of incomprehensible powers. Although we know that to dust we return and from dust we are made, the respective processes usually take long periods. But, at least for me, the emotions that accompany the stories differ in fundamental ways. Whereas I find the journey of Lot's wife only uncanny, I find the journey of Galatea both uncanny *and* beautiful, so much so, that the eeriness of watching a piece of stone turn into a woman is overcome by the feeling of joy upon watching creation at work. I imagine myself witnessing the creation of a new bio- and nootemporal reality, with all its unresolvable conflicts, from the relative chaos of inanimate stone. Since both examples employ a great degree of journeying among the temporal Umwelts, I will have to attribute the beauty of Galatea not the freedom of time-roaming but to the direction of the journey.

Art may praise both women, but I greatly prefer a woman named Galatea to a salt pillar named "Lot's wife." Of course, at the origin of all Galateas there is a salt statue; our primordial roots survive inside us, ready to tell us about their affinities to the primitive moods of time. If we follow them far enough we reach the emptiness and chaos that correspond to the atemporal world, from which inanimate matter emerged eons ago. The vistas, then, are those of the abyss. It is the office of the beautiful to imply the journey away from, rather than toward the abyss.

But how may this be done? With succinctness, Nikos Kazantzakis asked this question and gave an answer:

> How can you reach the womb of the Abyss to make it fruitful? This cannot be expressed, cannot be narrowed into words, cannot be subjected to laws; every man is completely free and has his own special liberation. No form of instruction exists, no Savior exists to open up the road.

Instructions about what is true and what is good are, indeed, irrelevant to this enterprise. They cannot guide us, because bringing about what is beautiful, that which makes the atemporal abyss fruitful, is the making of a new reality. Each aspect of the world so made is yet another journey of Galatea. The beautiful in thought, action, and feeling reveals, to maker and beholder alike, the possibility of man's participation in the cosmic process that creates conflict out of chaos.

XII Problems in the Study of Time

THERE ARE, PERHAPS, ONLY TWO WAYS in which very concise accounts of vast and complex phenomena may be given: through mathematical physics and through mysticism.

Thus, in Hinduism, OM is a sacred syllable of mystical potency: it contains the essence of the entire universe. Of course, it takes many years to learn why this is so and how this syllable is to be used in meditation. In the mathematical theory of relativity $R\mu\nu = O$ is a famous formula of great potency: from it follow all statements one can make about the motion of galaxies, stars, and in principle, of all particles. Of course, it takes many years to learn why this is so and how this formula is to be used in understanding the world.

The theory of time as conflict cannot claim the all-encompassing totality of mysticism or boast the pithy cogency of physical formulas, though it has something to say about both. As a theory in natural philosophy, it is an attempt to measure up the limits of man against the order of things by means of methods that satisfy speculative curiosity. As a class of principles in the scientific and humanistic study of time, it is not suited for abbreviated representation. This book itself is a summary.

Instead of attempting to abstract a summary, I have compiled a *Problemata,* which is a set of questions that may be used to explore the study of time in general, and the theory of time as conflict, in particular. They are loosely grouped under twenty-two lighthearted headings. All the problems pertain to material that has been dealt with earlier, but the chapter is not a map of this book.

Few, if any, of the questions can be answered with a simple true or false statement; they are more like themes for essays and treatises. Some have veiled or obvious suggestions; many reflect doubts; a few reflect certainties.

As a set of problems, this chapter may assist in the design of courses in the study of time. Many of the questions were developed in the seminars that I have conducted during the last fifteen years. Taken together, they demonstrate the intellectual wealth that resides in the serious study of time. They remind us that "a man's reach should exceed his grasp, or what's a heaven for?"

1   Clocks

What, precisely, do we mean by measuring time?

What are the conceptual and practical requirements which must be put to a machine or process before it may be admitted as a clock?

Time measurement necessarily involves at least two processes. "At the tone the time will be exactly 7:00 o'clock." What are the two processes?

Why do we describe the temporal organization of human acts in terms of suns, moons, and seasons (days, months, and years) and not the heavenly cycles in terms of human schedules?

A dial type wristwatch is a Platonic device, a digital clock an Aristotelian one. Are they totally interchangeable?

How do clocks and calendars reflect the prevailing judgments of a place and epoch as to what makes the universe tick?

Design a small clock. Then a smaller one. What will determine the size of the smallest useful clock?

I claim to have a perfect clock. What proof will you ask for?

"The more perfect the instrument as the measurer of time, the more completely does it conceal time's arrow." (Eddington) Why?

Timesect a working kitchen clock. Remove long hand, note new period (12 hours); remove short hands, shafts, gears. Change tools when you reach the balance wheel so that you may examine the structure of the metal. Keep a record of the cyclicities down to the atomic level. Where was the clockness of the clock hidden?

Professor Simplicio says he can reverse time. The Grand

Duke of Toscana arranged for you to visit Simplicio and ask for a demonstration. What kind of proof will you seek?

2   Rapid Transit

In a universe of pure light an (imaginary) observer must travel on a photon. What is the color of his universe?

In its proper frame the photon is atemporal, and hence it must fill all points along its path simultaneously, even from the beginning of the universe to its end. But the universe is expanding. What happens to the limiting frequency of photons, as seen by nootemporal observers?

Compare the ideas "relative motion faster than light" and "relative motion slower than rest."

The motion of light is to the idea of absolute motion as inertial translation is to the idea of absolute rest. With respect to which absolute do you prefer to formulate your kinematics and dynamics, and why?

Neglecting galaxies, what objects in nature travel at relativistic speeds?

Clock leaves home at 7:00 A.M. returns at 7:00 P.M. Let it demonstrate a special relativistic time dilation of one hour. Are the difficulties technical, or are we pushing against a conspiracy in nature?

What are the biological assumptions in the relativistic idea of differential aging? Psychological and religious implications?

Newtonian kinematics separates the lawful from the contingent in motion in a certain way; relativistic kinematics does it differently. What might be the future fate of the idea of space-time?

3   From Atoms to Thomas Jefferson

How can a set be empty and still be a set?

What evidence do we have from neurology, psychology, and physics for a smallest unit of time?

Do all chronons contain the same or different numbers of (mathematical) instants? Are chronons infinitely divisible?

An atomic chronon is as atemporal as is the absolute elsewhere of space-time. Are the two regions contiguous?

Put two chronons end-to-end. What have you got?

What are the similarities and differences between spatial and temporal atoms?

If there were such things as atoms of time (chronons), how would you combine some of them to make up a night at the opera?

How are atomic and perceptual chronons entered on the Minkowski diagram?

Aphrodite and Pan were throwing dice, each with six, numbered, sides. Eros improved the dice until they became ideally true. What happened to the game?

It has often been claimed that Boltzmann's H-theorem connects the timeless world of particles with the macroscopic world of time. Does it?

Define futurity and pastness without reference to a present.

Diadic time means an earlier/later relationship without a "now." That is, event B is in the future of event A, and event A is in the past of event B. Is there not a present hidden, somewhere?

Can a temporal event turn into a spatial thing, or vice versa?

It has often been stressed that time and space enter four-space on equal footing. But special relativistic time dilation is real in the sense that it is permanent, whereas rod contraction is virtual in the sense that it is impermanent. Whence the asymmetry?

In physics, t usually stands for "time." It has been assumed that a t is a t is a t. Is it?

What in nature corresponds to that sharp contrast between time and the timeless which we ordinarily hold to be associated with a beginning or an ending of time?

Do motion and rest form a true, mutually exclusive disjunction?

What conditions correspond to ideas of beginnings and endings on different integrative levels of nature?

Change, we say, takes place in the course of time. What may be meant by the idea that, in the course of evolution, new temporalities arise?

It is the case that Thomas Jefferson is not now alive. For this, and for many other reasons, I agree with the Bard that

> Like as the waves make towards the pebbled shore,
> So do our minutes hasten to their end.

For me, time passes. Does it pass for horses? For fleas? For DNA molecules? For the moons of Jupiter? For hot plasma? For quanta of light blue light?

4   Jails and Arrows

Design a box that corresponds to the idea of a container which may be labeled "Closed System."

Put a physicist in a closed system so that he will be part of the processes going on therein. What will he observe, and what languages must he use, as the contents of the box reaches a state of maximum entropy?

All closed systems change so that, on the average, their entropy increases. Imagine the net increase as the difference between two, opposing, semi-infinite arrows. What are the opposing trends represented by the arrows? What will be the state of the system when the opposing trends vanish?

Certain physical principles minimize the rates of entropy increase of closed systems. Living organisms can do better, at least locally. In what way?

The sun, a pasture, and a horse make up a closed system. On the average, the entropy of the system increases. On the average, the entropy of the living horse decreases. The arrow of time parallels which of these two systematic changes?

5   Life and Sex

I have just created life from inanimate matter. By what criteria will you judge the validity of my claim?

I have constructed a molecular clock that models the tidal rhythm. In a protected, collaborating environment it grows like crystals do. When it is too large, it splits into little crystals. Please test it to see whether it is alive.

Timesect a living organism. Tabulate its cyclic spectrum from slow adaptive changes (frequencies) up to light sensitivity. Does the spectrum suggest a master clock, or is the organism more like a coordinated clock shop?

Adaptation by a broadening spectrum of cyclic functions forces upon the organism a division of labor by frequency. Correlate the portions of the spectrum with appropriate physiological clocks. What are the upper and lower limits of these frequencies?

Construct a quantitative index that corresponds to the idea "the difference between the expected and the encountered by the organism." (For man, call it "the difference between hope and reality.") Describe this difference as existential tension. Test its variations with complexification and with improved Darwinian fitness.

The necessary inner coordination of living organisms inserts a meaningful present in the pure succession of the eotemporal world. Why, then, is your now the same as mine? Or, is it?

I died from mineral and plant became,
Died from the plant and took a sentient frame.
Died from the beast and donned a human dress,
When by my dying did I e'er grow less?

(Jalal-ud-din Rumi)

How did the aging order of life evolve from the cyclic matrix?

Does a molecule age? Can a hydrogen molecule determine a present? The DNA is a molecule; can it determine a present? Do DNA molecules age?

Mammalian germ cells divide, but upon division, no dead bodies remain. Are mammalian germ cells immortal?

What, precisely, are the spatial limits of the effectiveness of man's genes?

If each new soma is but another experiment by the gene, where does human responsibility enter?

Is the capacity to reproduce a necessary feature of life?

What are the advantages/disadvantages of that periodic regression from phenotype to genotype which is characteristic of advanced life forms?

Compare the rates at which mutations may spread in a population that reproduces asexually with the rates for one that reproduces sexually.

Why is the life cycle the basic unit of evolution?

What might have been Plato's opinion of the Christian doctrine of the First Intercourse (Original Sin)?

What might have been St. Paul's opinion of the idea that aging and death are complementary forms of adaptation?

> A woman can be proud and stiff
> When on love intent;
> But Love has pitched his mansion in
> The place of excrement;
> For nothing can be sole or whole
> That has not been rent.
> (W. B. Yeats)

What evidence is there on the common evolutionary origins of sexual reproduction and death by aging?

6 Evolution

Evolutionary adaptation is "adjustment to environmental conditions by organisms or a population so that it becomes more fit for existence under the prevailing conditions." Do environmental conditions also evolve by adaptation? If not, why not? If yes, what does the adapting: the monkey or the banana?

How does the issue of time available versus time necessary enter the arguments of teleology in evolution?

Assume that organic evolution favors the fastest adaptive measures. How could the effects of such a policy be distinguished from goal-directedness in nature?

The most perfect adaptation is by death. In a while, the body will so superbly fit for existence under the prevailing conditions that it will be indistinguishable from the environment. Why, then, do living things, individually and collectively, adapt through survival rather than through death?

The evolution of knowledge (noetic learning) is fast. Genetic learning is slower, the evolutionary rate of the earth is even slower. The evolution of the universe is the slowest. Put these statements in forms suitable for quantitative testing.

Growth and decay are always simultaneously present in all living organisms. Why do we usually identify life with growth rather than with both?

What happens to life when the opposition between growth and decay vanishes?

The necessary decrease of entropy of living systems is an insignificant (statistical, local) variation, from the point of view of the second law of thermodynamics. How significant or insignificant is it from the point of view of the organism?

Why do old Chinese magnetic needles point south, whereas old Western magnetic needles point north?

Both the entropy growth and the entropy decrease arrows of closed and open systems, respectively, parallel our sense of time. Why does the received teaching of physics insist that time's arrow is along the increasing entropy mode of the inanimate world?

7   Memories and Hopes

Grey squirrels begin building their winter nests three months before they are needed. People begin saving for the college education of their children 120 months before the funds are needed. In what ways are these anticipatory actions fundamentally different?

"Nessun maggior dolore che ricordarsi del tempo felice nella miseria." (Italian saying) If memory corresponds to a brain state, ideas about the future to another, present sense impressions yet to another, and if they are all simultaneously present, how do we tell them apart?

Make yourself remember the graceful motion of a friend. Do this without forcing upon the images your present awareness of passing time. Do you see motion or a static tableau?

Consider the capacities of appreciating the future and the past as two distinct categories of time. Add to them the ability of man to combine memories, expectations, and sense impressions in a mental present. In what sequence do these capacities develop in the mind of the growing child? In what sequence have they come to be expressed during the growth of civilizations? In the course of organic evolution?

Trace the earliest memories of your childhood. Surely, your genetic endowment carried some memories older than the ones you remember. Where might be found the ultimate roots of memory?

What is the difference between mortal terror and the fear of death? Would it assist the survival of our species if we knew exactly how and when we should die?

Classical empiricism held that the sources of all knowledge were in the experience of the individual. Reformulate this principle in post-Darwinian terms.

8   Your Self

If skilled surgeons were able to interchange the heads of two healthy adults, would the heads brag about new bodies or the bodies flaunt their new heads?

How do you know that the body that fell asleep last night is the same that awoke this morning?

Who or what is the "I" in the shout "I don't want to die" coming from a man standing before a firing squad?

Who or what is the "I" in the whisper "I don't want to die" coming from a man with cancer throughout his body?

Who or what is the "I" in the sentence "I am happy to die" coming from a man about to sacrifice himself for a cause?

Why is it so difficult to share the concerns of a cockroach?

According to St. John "In the beginning was the Word." What is the privileged position of language in the development of identity?

To what extent is human language an individual enterprise, to what extent is it a collective one?

Separate the emotive from the cognitive in your experience and communicate their distinctness through other than the spoken or written tongue.

I cannot explore myself by sight as I can explore others. Can I do so by touch? By hearing? By smell?

Describe your identity through means other than the spoken or written language. Music? Dance? Painting? Barking?

What might possibly be meant by natural selection among "brain children" (ideas)?

Man's awareness of the passing time seems to have an irreducible component of tension. It might be identified with the difference between hope and reality (the expected and the encountered). What happens to our sense of time if the expected and the encountered coincide in all details, for an appreciable period of time?

Would a creature without anxiety explore the past? The future? How does nature protect the unresolvable conflicts of the mind, that is, the integrity of the self?

How may past and future look to a dying man?

Imagine your death: your body is still, your relatives crying, your soul elsewhere. But after you die, there will be no self to observe your dead body. There is something strange about the idea of imagining your death. What is it?

State the classic debate of free will versus determinism in terms of relativistic space-time.

Would it be beneficial to you and/or to the survival of our

species if you could judge your actions with complete detachment?

> Seltsam in Nebel zu wandern!
> Leben ist Einsamsein.
> Kein Mensch kennt den andern,
> Jeder ist allein.
>
> (Hermann Hesse)

What is the usefulness of selfhood for the survival of the species?

9   Your Brain

The difference in mental capacity between man and the most advanced primates is so vast that the differences in weight, volume, or size of the brain (or prorated weights, volumes, or sizes) give no clues as to the source of that difference. What, then, makes the human brain unique?

During organic evolution, the humanoid features of the brain seem to have emerged with striking rapidity. What may account for this economy of time?

Let complexity be defined as the number of possible states involving all members of a coherent unit: a riding club, a telephone network, a brain. It has been calculated that the number of possible brain states is perhaps 10^{10^9}. The brain, according to our definition, is very complex. Find examples of similar complexities.

The very large, the very small, the very cold, the very hot, the very fast, and the very many all have their peculiar laws, unpredictable from the laws of the average size, speed, etc. What peculiar laws might control the very complex?

The problems of this chapter give a process description of my brain: an account that unfolds in time. I could not give a state description of my own, or of any other brains. Whence the asymmetry of usefulness between process and state descriptions for this particular task?

10   Time Passes

"And God had him die for a hundred years and then revived him and said: 'How long have you been here?' 'A day or a part of the day,' he said." (The Koran) What is the prevalent temporality of the manifest content of dreams?

Why is it difficult to integrate alleged instances of precognition in the structure of natural science?

"What is six winters? They are quickly gone. To men in joy; but grief makes one hour ten." (Shakespeare) What controls the apparent speed at which time passes?

Along the hierarchy of integrative levels, where do the functions of the Id belong?

Why is it that in childhood and youth time, generally, passes slowly; as we age, it passes faster and faster?

Why is it that "One crowded hour of glorious life is worth an age without a name?" (Thomas Mourdant)

11   The Universe

At various epochs and places, the universe has always been judged best describable by whatever happened to have been the favored mode of explanation. Does this say anything about the universe? About man? About both?

Take four balls and construct a system wherein they are mutually equidistant. Take a fifth ball and make a new system wherein all five balls are mutually equidistant. If you don't succeed expand your model to astronomical size and measure distances with lasers and clocks. What happens to your geometry?

Models of the universe which use astral geometry employ a single variable: cosmological common time. But if the equations say naught about futurity, pastness, and presentness, whence the temporal passage?

The metaphors "the universe expands" and "the universe flattens out" are equivalent to "the geometry of the universe is time-dependent." But there is nothing in physics that could determine a now, a future, and a past. How has time been smuggled in?

The universe of astral geometry is said to be unbounded but finite. Could it also be described as infinite but finite? Bounded but unbounded?

For Cusanus, the finity-infinity of the world became acceptable once he accomodated it to the idea of God. In scientific cosmology the finity-infinity of the world became acceptable through the teachings of astral geometry. Compare the assumptions implicit in each argument.

We do not live at the center of the earth; the earth is not the center of the solar system; the solar system is not the center of our galaxy. But we are at the center of the universe of scientific cosmology. Where, in the course of the arguments, did our perspectives change?

The finite but unbounded universe of four-space includes all galaxies, stars, planets, boulders, squirrels, monkeys, you and me, and the memory of George Bernard Shaw. Does it?

By definition, the universe is everything there is. What may be meant by an ensemble of universes?

What is the temporality of the topologically closed time of scientific cosmology?

If eotemporality is the highest temporal level of the physical world, how can one tell whether the beginning of the universe (if there was such an event) is in the past or in the future?

"Brother Fire, before all things, the Most Holy has created Thee of exceeding comeliness." (St. Francis of Assisi) What kind of clocks can we use to measure time during the first few microseconds after the expansion of the primordial fireball began?

It is now 1978 A.D. Pack a suitcase of clocks to be used in your journey back to the singularity of physical cosmology. What will you take?

According to scientific cosmology the Big Bang happened about 5×10^{17} seconds ago. According to Bishop Ussher the world was created in 4004 B.C. The estimates differ by a scale factor of 10^6. What do they agree upon?

An astronomer has just photographed the edge of the universe which, assuming that the universe expands, is the same as photographing the beginning of time. What, do you suppose, the photograph shows?

An artist has just drawn a picture of the universe as it may look when time ends. What details will you examine so as to assure yourself of the reasonableness of the artist's claim, or else reject it as mere fancy?

12   The Universe of Man

"Progress is our most important product." (Advertising slogan) When and how was the idea of linear history born?

How can uniformitarianism accomodate nomogenesis?

What are the relative predictive powers of astronomy, geology, organic evolution, history, and sociology?

Instructions for conduct may be easily tied to universal cosmologies because they admit a mutuality between man and the world. It is very difficult to tie ethical standards to scientific cosmology. What has been gained? What was lost?

List some ways in which time budgets depend on the ethnic and cultural background of the individual.

Visit an ancient pharaoh whose slaves are building a pyramid. Explain to him the Marxist interpretation of history or the economics of Adam Smith. Assume that he is bright and well-informed. What might he say? (Hint: get ready to run!)

Scientific laws, unlike religious principles, must be testable. Against what other universes should we test the laws of scientific cosmology?

"Basic group identity ... is a living thing that grows, changes, and thrives according to the rise and decline of its own vitality and the conditions in which it exists." (H. R. Isaacs) Try to formulate a single, foolproof criterion whereby an American may be identified. A boy scout? A woman? A candlestick maker?

"One nation, indivisible, under God." This description makes sense only if certain assumptions are made about the rest of the world. What are these assumptions?

Against what other units of people might we test the

validity of economic and social theories postulated to hold for a world-wide society?

13   Unstable Conditions

Try to identify a physical state in which energy is between its radiative and particulate forms.

If life had really come about from non-life, why cannot we find emergent life in our own epoch?

Is a DNA molecule inanimate or is it alive?

Why are there no creatures today occupying an unambiguous position between man and ape?

Why do many people consider psychiatry as a profession between that of a physician and that of a minister?

What policies (laws, regularities) of nature help keep the major integrative levels distinct? What policies provide continuity between adjacent levels?

What difficulties did Franz Kafka's *K* encounter when he tried to cross the language boundaries of the Castle?

Nature's languages (laws) are level-specific. Each language subsumes those underneath it and adds to them some of its own peculiarities. What does this condition presage about the success of an attempt, designed to derive the instructions for curing ham from what we know about the hydrogen atom?

How is continuity retained between the growth/decay conflict of living matter and the roots of this conflict in inanimate matter?

What constitutes the fiasco of the conflict resolution possible for inanimate matter? How is this limitation overcome?

What constitutes the fiasco of the conflict resolution possible for living matter, and how is this limitation overcome?

What constitutes the fiasco of conflict resolution possible for the individual mind?

Why is it that catastrophe theory, in its present form, is not suited to handle the mathematics of transitions between integrative levels?

Biogenesis probably preceded the evolution of the capacity to produce offsprings. What selective advantages might have led to the emergence of reproduction?

The evolution of the brain surely preceded the human capacity to create symbolic transformations of experience. What selective advantages might have led to the emergence of the mind?

If the past is a reliable guide, what is likely to happen to man on earth when the network of signs and symbols, created by the community of minds, crosses a certain threshold of complexity?

14   Modes of Knowledge

To support a weight one must stand under it. To support an argument one must understand it. What is biosemiotics?

The taste of a kiss: a very intimate experience. The taste in tractor design: it is there for all to see. What is common to these disparate meanings of "taste"?

A man grasps a woman and makes her conceive new life. A man grasps the significance of a formula and explains it to his class. What is common to these two meanings of "grasp"? Do you see what I mean? Why do lovers, scientists, and judges seek to discover the naked truth? What is truth?

"The forms of our limbs are made in similitude to signs of hidden higher things that the intellect can only comprehend by way of mnemonic comparisons." (Joseph Gikatila) Is this biosemiotics? Cabala? Nonsense?

Light is conceived of in all great cosmologies as a symbol and source of reason. Ignorance is represented as darkness. Why not sound versus silence, roses versus manure, or heat versus cold?

A young man may feel that he has been called to serve God and nation. Why not signaled (as by a lamp) or seduced (as by a pheromone)?

Damnyankees may never be decent, damn Yankees might be. What comes first: the word or the thought?

The child learns words by repeating them. He hears sentences. Then he constructs sentences he had never heard before. Whence this very useful capacity?

When is a metaphor a metaphor? When is it an explanation? When a theory?

15 Quantity and Quality

Is "one" a number?

"One, two, button my shoe; three, four, shut the door." Which does the child learn first: cardinality or ordinality?

Compare the developmental stages of the child's ability to handle numbers with stages in the history of mathematical knowledge, then, both with the types of mathematics appropriate for the hierarchy of temporalities.

The Augustinian indeterminacy principle: if no one asks me what time is, I know what it is; if I wish to explain, I know not. In what way did modern science alter the complementary qualities of time felt and time understood?

"I felt before I thought: this is the common lot of humanity." (Rousseau) As an individual, how do you like to play the balance between your archaic paleologic and the newer, discursive logic? Is your occupation consistent with this preference? What are your favorite metaphors of time?

What are the reasons for the tremendous effectiveness of mathematics in natural science?

It has been estimated that in 1978 perhaps three million little Americans will be born. What does this tell me about the family life of Uncle Alan and Aunt Pat?

Why does the effectiveness of number drop so rapidly when one leaves the inanimate world and begins considering the important questions of life, mind, ethics, aesthetics?

16 Science and Truth

Until about the 14th century the Chinese showed a clear

technical superiority to the West. They also mastered mathematics and had a profound interest in nature. Why did the scientific revolution take place in the West and not in China?

"The Christian knows," wrote Kepler, "that the mathematized principles according to which the corporeal world was to be created are coeternal with God." How does the Christian know that?

The industrialized West is fond of things that work. It is concerned with outer space (astronomy, chemistry, the life and mind of others). It prefers the predictable, the mathematical, the analytic. Make a self-consistent list of opposite preferences. Now, imagine a civilization that is fond of these (new) values. What might be their criteria for truth?

Spend five years each in central Italy, Western Norway, North-West India, central China. Weigh relative preferences for the quantified versus the organic type of knowledge. Compare attitudes toward earthly possessions; compare sexual mores. Compare their expectations about the future well-being of their citizens.

Study extinct civilizations through records and artifacts. Visit ruins. Ask yourself the same questions as those above.

Modern prophets make numerical predictions based on quantitative laws. Ancient prophets made qualitative predictions based on beliefs about man and God. What has been gained? What has been lost?

Why has the use of number magic been so effective in advertising?

For universal acceptability, quantified knowledge has great advantages over religions and social doctrines. Should the true and final Utopia of mankind be governed by scientific knowledge? Social doctrine? Religious revelation? All or some of these?

Are testable predictions necessary for the validity of a principle in natural science?

"But let Time's news be known when 'tis brought forth." (Shakespeare) Should time be identified with being? Becoming? Both? Neither?

Consider some formulas of physics, some laws of chemistry, and a number of well-established principles in biology. Try to formulate a general statement that will tell us where lawfulness stops and the generative aspects of time begin, as seen in natural science.

"Truth is that class of human knowledge which individual and communal judgments regard as permanent." Does it make sense to transpose this definition to the species-specific worlds of animals?

When I was a child the wind blew only when the trees stirred up the air; the snow kept the winters cold until, finally, the snow decided to melt. What are the difficulties of a causal theory of time?

17   Praiseworthy Conduct

Can a particular crocodile be said to be good or evil? A particular human deed? A particular man?

"Joseph also went up from Galilee ... unto the City of David." Why did he go "up"?

Compare the impotent rage of the infant with the impotent rage of the dying man.

There is evidence that Paleolithic man was already concerned with post-mortem survival. Why have people been so much interested in life after death?

Why are so many people concerned with the issue of nonbeing after death, and so few people with nonbeing before conception?

Why is it that, to the extent that religious teachings influence behavior, they do so through eschatologies rather than cosmogonies?

Experiment by observing your attitudes toward your own powers to sway the future from evil to good. (i) Join a litany procession before Ascension Day and, together with others, invoke the blessings of a favorable season. (ii) Visit the temples of Kyoto on the day of the vernal equinox, ring the bells, burn incense, and invoke same.

What the New York subway system misses is the integrated tranquility of Tao. What are the ethical paradigms appropriate to the two, distinct, atmospheres?

Why is it that watered-down versions of Buddhism and Hinduism have such great appeal to the rootless of modern America?

How does gnosticism enter cybernetics?

What do the sciences teach about good and evil?

"A greater power than we can contradict hath thwarted our intents." (Shakespeare) History is, of course, very complicated. But if you were hard pressed to find some lawfulness to history, where would you look?

Are ethical teachings discoveries of preexisting truths or determinations of yet undetermined futures?

"The peace of God that passeth all understanding." (St. Paul) Compare the timelessness of the ecstasy of the chalice with the atemporality of relativistic gas.

Why does moral guidance tend to have an aura of coming from "beyond and above?"

Why is obscurantism so often invoked as a proof of access to higher truths?

We protect our children. This is life, protecting life. Why do we defend our "brain children?"

"If all else fails, myself have power to die." (Shakespeare) Does man's capacity for suicide demonstrate the existence of free choice?

In suicide, who is the predator and who is the prey?

"My future will not copy fair my past." (Elizabeth Barret Browning) Given the strict predictability guaranteed by the laws of macroscopic matter, can this claim have any truth value?

18   The State

Compare the ideals of the City (meaning, society) envisioned by Plato, St. Augustine, Machiavelli, Marx, and your favorite futurologist.

Bare ruin'd choirs, where late the sweet birds sang.
In me thou see'st the twilight of such day
As after sunset fadeth in the west.
                                        (Shakespeare)
Phrase a question appropriate to the quote as well as to this section.

The discovery of the laws of planetary orbits did not change those orbits. The formulation of Christian ethics did alter the fate of the faithful. What is the epistemic status of ethical teachings?

Select a community governed by priests, another by bankers, one by atomic physicists, one by merchants. Compare the prevailing standards of praiseworthy conduct.

Did the women's movement succeed in inserting the feminine among the paradigms of praiseworthy conduct?

What is the difference between altruism and compassion?

What would human life be like in a society whose members, within each profession, are indistinguishable and hence, interchangeable?

Why is it that if only one lifeboat is available, women and children should go first?

Prophets, thinkers, and poets of all ages have advocated societal adjustments which did not lead to equilibrium by way of redress, but produced disequilibria in need of redress. All great social upheavals of history are of this nature. Can this fact be reconciled with the interpretation of behavior as a means of adaptation?

Queen bees kill their fertile daughters. Is this their duty? Are they committing a crime? Would they commit a crime by not committing murder?

What is the evolutionary office of moral responsibility?

What pattern of behavior may be associated with the idea of the totalitarian mind?

"The prophet that hath a dream let him tell the dream." (Jeremiah) What is the likely fate of a nation whose prophets are silent?

"'Tis the final conflict, let each stand his place, the International Party shall be the human race." (Eugene Pottier) Can individuals and collectives remain human in the absence of certain unreducible conflicts?

"Time's glory is to calm contending kings." (Shakespeare) If the world had but a single king, what would this do to time's glory?

19 People on Earth

What advantages accrue to industry from minimizing sex-coded roles? What advantages for a totalitarian government?

The effectiveness of birth control will surely lead to the societal acceptance of two distinct functions which may, but need not, coincide: falling in love and falling in sex. What is the likely fate of the family?

> Who goeth a borrowing
> Goeth a sorrowing
> Few lend (but fools)
> Their working tools.
>
> (Medieval doggerel)

What did exosomatic evolution do to our production methods? To our cosmologies?

Membership in a human anthill of indistinguishable individuals forces the mind to operate in a prototemporal Umwelt. What psychological consequences may be predicted?

Genetic learning is much slower than noetic learning. Make reasonable assumptions about their comparative speeds, then estimate how many people must inherit the earth for genetic learning rates to catch up with noetic learning rates?

How does Marxist thought interpret the role, causes, and purpose (if any) of suffering in human life?

Draft the outlines of death-control legislation for the United States Congress. Do the same for the Supreme Soviet.

The 200,000 new mouths to be fed daily exemplify the

madness of the wildly reproducing biomass of mankind. Develop
plans that would assist in the production of material goods and
health care for the increasing populace yet safeguard the dignity
of man.

What do Capitalism and Communism teach about the
dignity of the individual?

Marry several husbands and love them equally. Lose one to
the sea; one, as a nameless member, to a Russian concentration
camp; one to cancer; one to an insane asylum; one to a youth
gang in New York City. Based on your experiences, describe the
kind of society in which you would like to live.

Is it possible to accomodate cultural religious, linguistic,
racial, etc. pluralism and also maintain the homogeneity necessary
for the running of a modern industrial state?

Search history for cues: what are the likely effects of a
summary change in our calendrical system? Of a slow change?

If today we have conditions without historical precedence,
of what possible use is the study of history?

According to Jorge Luis Borges it is not reading but
rereading that matters. If this is so, what is the presumptive fate of
books in a society which identifies time with continuous change?

Write a play or a novel about a child that grew up happily
because his father subscribed to the Toy of the Month Club.

> My mummy is a picture tube;
> I watch her 'til I'm dizzy!
> She is better than my real one is,
> 'Cause that one's always busy.
>
> (Mad Magazine)

What are the advantages/disadvantages of machines as substitute
mothers? Lovers? Enemies?

In *The Lord of the Rings*, J.R.R. Tolkien created a new
language. In English it is called the Westron. Construct a language that can express only economic and production instructions
in form of imperatives, but excludes the possibility of expressing
speculative curiosity. What kind of temporality would be suitable
for such a language?

The Malthusian principle of population is based on the differential growth rates between human population and food. Extend this principle to crises produced by other growth differentials: the demands of the human mind outrunning the capacities of the human body; the demands of a collective of minds outrunning the capacity of the individual mind.

20   Temporal Moods

Upon critical examination, the ordinary use of the idea of "timelessness" proves to be confused. Clear up the confusion.

Compare, and distinguish among, the timeless ecstasies of the dance, the forest, the bower, and the chalice.

What are the political uses of timeless ecstasies?

The affective dimensions of the temporal Umwelts are their "moods." What is the dominant temporal mood of dreams? Of cubism? Of surrealism? Of "socialist realism?"

Why are most paintings of Hieronymous Bosch uncanny?

"It was a strange jargon – the Lord's Prayer repeated backwards – the incantation usual in the proceedings for obtaining unhallowed assistance against an enemy." (Thomas Hardy) What's wrong with things going backward? Why is it that watching almost any movie run backward, is funny?

Why is it that experiences of *déjà vu* and *jamais vu* are usually uncanny?

Time travel is an intellectually dishonest enterprise. Why is it so popular?

What are the taboos against tempering with temporal categories? What do these taboos protect?

21   Galatea and Lot's Wife

Variations in aesthetic judgments are vast. Does it make sense to claim universality for any statement one can make about the beautiful?

Why are so many birds, butterflies, and deep-sea fish so beautifully colored?

Separate, if possible, beauty of appreciation from beauty of usefulness for (i) flowers, (ii) birds, (iii) birdsongs, and (iv) women.

Scientific creativity discovers details of the world that had remained unknown until they were discovered. How about artistic creativity?

In what fundamental ways does music differ from speech?

Only the unpredictable part of language carries information. The predictable we already know. Why do we reread good literature and revisit beautiful buildings?

Mathematics is a suitable language for a noetic cosmology. What languages are best suited for the writing of cosmologies of emotion?

Calculate the frequency spectrum of Mozart's *The Magic Flute*, as a unit of repeatable sound. How does its range compare with that of the cyclic order of life?

A playwright writes plays. A filmwright may be a director, a cameraman, script writer, actor – or all of them together. Compare the freedoms of playwrights and filmwrights in handling temporal experience.

How do comedy and tragedy differ in their temporal perspectives?

Tragic destiny is an unpopular idea in our epoch. Tragedy as a form of drama is all but taboo on the stage. Why?

What circumstances tend to change tragedy into comedy and vice versa?

Can a single instant of life look comical? Horrifying? Tragic? Sad?

Time as a commodity may be bought or sold. Chickens as a commodity may be bought and sold. What does this happy kinship between time and chickens do to the idea of salvation history? To our ideals of compassion? To heroism, as an ideal?

Lot's wife collapsed from a living woman to an amorphous rock of salt. Galatea emerged from an inanimate rock and took her place among living, struggling, feeling women. Compare their journey along the temporal Umwelts.

22   Have We Said Anything?

It has often been claimed that whereas the sciences are difficult to master, the humanities are relatively easy, because they deal with the universal experience of being human. True? False? Impossible to decide?

We live in an epoch of proliferating knowledge and with a cacophony articulated emotions. Can we assume that the informed mind is capable of erecting universal structures of thought today, as it is known to have done in the past?

> Prince, n'enquerez de sepmaine
> Où elles sont, ne de cest an,
> Que ce refrain ne vous remaine:
> Mais où sont les neiges d'antan?
>   (Villon)

After attending to almost three-hundred problems, I give up. Where are the snows of yesteryear?

> We have given you a world as contradictory
> As a female, as cabbalistic as the male,
> A conscienceless hermaphrodite who plays
> Heaven off against hell, hell off against heaven,
> Revolving in the ballroom of the skies
> Glittering with conflict as with diamonds:
> When all you ask us for, is cause and effect!
>   (Christopher Frye)

What comprises the theory of time as a hierarchy of unresolvable, creative conflicts?

New Gilead, Connecticut
September 10, 1977

General Sources Consulted

Bonner, J.T., (1974) *On Development: the Biology of Form*, (Cambridge: Harvard University Press).
Brandon, S.G.F., (1962) *Man and His Destiny in the Great Religions*, (Manchester: Manchester University Press).
–, (1965) *History, Time and Deity*, (Manchester: Manchester University Press).
Davies, P.C.W., (1974) *The Physics of Time Asymmetry*, (Berkeley: University of California Press).
Doob, L.W., (1971) *Patterning of Time*, (New Haven: Yale University Press).
Fischer, Roland, (1967) *ed.,* "Interdisciplinary Perspectives of Time," *New York Academy of Sciences, Annals*, v. 138, art. 2, p. 367–915.
Fraser, J.T., (1966) *ed., The Voices of Time*, (New York: Braziller).
–, (1975) *Of Time, Passion and Knowledge: Reflections on the Strategy of Existence*, (New York: Braziller).
–, F.C. Haber and G.H. Müller, (1972) *eds., The Study of Time I*, (New York and Heidelberg: Springer Verlag).
–, and Nathaniel Lawrence, (1975) *eds., The Study of Time II*, (New York and Heidelberg: Springer Verlag).
–, Nathaniel Lawrence and D.A. Park, (1978) *eds., The Study of Time III*, (New York and Heidelberg: Springer Verlag).
Georgescu-Roegen, Nicholas, (1971) *The Entropy Law and the Economic Process*, (Cambridge: Harvard University Press).
Gunnell, J.G., (1968) *Political Philosophy and Time*, (Middletown, Conn: Wesleyan University Press).
Leon-Portilla, Miguel, (1973) *Time and Reality in the Thought of the Maya*, C.L. Boilès and Fernando Horcasitas, *trs.,* (Boston: Beacon Press).
Patrides, C.A., (1976) *ed., Aspects of Time*, (Manchester: Manchester University Press).
Scheving, L.E., Franz Halberg and J.E. Pauly, (1974) *eds., Chronobiology*, (Tokyo: Igaku Shoin).
Sherover, C.M., (1975) *The Human Experience of Timeï the Development of its Philosophic Meaning*, (New York: New York University Press).
Whitrow, G.J., (1961) *The Natural Philosophy of Time*, (London and Edinburgh: Nelson).

Literature Cited

Atkin, R.H., (1974) *Mathematical Structure in Human Affairs*, (New York: Crane and Russak).
Beckett, Samuel (1954) *Waiting for Godot*, (New York: Grove Press).
–, (1958) *Endgame*, (New York: Grove Press).
Beer, Stafford, (1967) *Decision and Control*, (New York: Wiley).

Bergman, Ingmar, (1960) *Four Screenplays of Ingmar Bergman,* (New York: Simon and Schuster).
–, (1972) *Persona and Shame,* Keith Bradfield, *tr.,* (New York: Grossman).
Bernal, J.D., (1967) *The Origins of Life,* (Cleveland: World).
Brain, Lord, (1963) "Some Reflections on Brain and Mind," *Brain,* v. 86, p. 381–402.
Browne, Thomas, (1658) *The Garden of Cyrus.*
Cairns-Smith, A.G., (1971) *The Life Puzzle: On Crystals and Organisms and on the Possibility of a Crystal as an Ancestor,* (Edinburgh: Oliver and Boyd).
–, (1975) "A Case for Alien Ancestry," *Proceedings of the Royal Society, London,* B.v. 189, p. 249–74.
Chaitin, G.J., (1975) "Randomness and Mathematical Proof," *Scientific American,* (May), p. 47–52.
Conrad, Joseph, (1897) *The Nigger of the Narcissus,* (New York: Collier, 1966).
Cusanus, Nicolaus, (1440) *Of Learned Ignorance,* Germain Heron, *tr.,* (London: Routledge & Kegan Paul, 1954).
Dostoyevski, F.M. (1864) *Notes from Underground,* B.G. Guerney, *tr., in* C. Neider, *ed., Short Novels of the Masters* (New York: Reinhart & Co., 1948).
Dürrenmatt, Friedrich, (1965) *Four Plays,* (New York: Grove Press).
Eddington, A.S., (1928) *The Nature of the Physical World* (Ann Arbor: University of Michigan Press, 1958).
Ehrenzweig, Anton, (1965) *The Psychoanalysis of Artistic Vision and Hearing,* (New York: Braziller).
Ehret, C.F., (1974) "The Sense of Time: Evidence for its Molecular Basis in the Eurkaryotic Gene-Action System," *Advances in Biological and Medical Physics,* J.H. Lawrence and J.G. Hamilton, *eds.,* (New York: Academic Press), v. 15.
Einstein, Albert, (1922) *Sidelights on Relativity,* G.B. Jeffrey and W. Perrett, *trs.,* (London, Methuen).
–, (1949) "Autobiographical Notes," *Albert Einstein: Philosopher-Scientist,* P.A. Schilpp, *ed.,* (New York: Tudor), p. 1–96
Encyclopedia of Bioethics, (forthcoming) W.T. Reich, *ed.,* (Glencoe: The Free Press).
English, H.B. and A.C. English, (1964) *A Comprehensive Dictionary of Psychological and Psychoanalytical Terms,* (New York: McKay).
Erikson, E.H. (1968) *Identity, Youth, and Crisis,* (New York: Norton).
Fowles, John, (1975) *The Ebony Tower,* (New York: The New American Library).
Fraser, J.T., (1970) "Time as a Hierarchy of Creative Conflicts," *Studium Generale,* v. 23, p. 597–689.
–, (1978) "Aspects of Time, Infinity, and the World in Enlightenment Thought." Read the Eighth Annual Meeting, *American Society for Eighteenth Century Studies.* Unpublished.
Galileo, Galilei, (1623) *Opere,* (Firenze, 1844), v. 4 "Ill Saggiatore." A translation may be found in *The Controversy on the Comets of 1618,* Stillman Drake and C.D. O'Malley, *trs.,* (Philadelphia: University of Pennsylvania Press, 1960).

Gheorgiu, V.C., (1950) *The Twenty-Fifth Hour*, R. Eldon, *tr.*, (Chicago: Henry Regnery).

Gibson, J.J., (1966) *The Senses Considered as Perceptual Systems*, (Boston: Houghton Mifflin).

Gödel, Kurt, (1962) *On Formally Undecidable Propositions of Principia Mathematica and Related Systems*, B. Meltzer, *tr.*, (Edinburgh and London: Oliver & Boyd).

Goodwin, Brian, (1976) "On some relationships between embryogenesis and cognition," *Theoria to Theory*, v. 10, p. 33–44.

Granet, Marcel, (1975) *Chinese Thought*, (New York: Arno Press).

Haldane, J.B.S., (1929) "The Origin of Life," reprinted in J.D. Bernal, *The Origins of Life*, (Cleveland: World, 1967), p. 242–9.

Hartocollis, Peter, (1976) "On the Experience of Time and Its Dynamics with Special Reference to Affects," *Journal of the American Psychoanalytic Association*, v. 24, p. 363–75.

Hastings, J.W. and Hans-George Schweiger, (1976) *eds., The Molecular Basis of Circadian Rhythms*, (Berlin: Abakon).

Illich, Ivan, (1976) *Medical Nemesis: The Expropriation of Health*, (New York: Pantheon).

Jerison, H.J., (1973) *Evolution of the Brain and Intelligence*, (New York: Academic Press).

Jones, M.R., (1976) "Time, Our Lost Dimension: Toward a New Theory of Perception, Attention, and Memory," *Psychological Review*, v. 83, p. 323–55.

Kazantzakis, Nikos, (1960) *The Saviors of God*, Kimon Friar, *tr.*, (New York: Simon and Shuster).

Kepler, Johannes, (1619) *De Harmonice Mundi* in *Gesammelte Werke*, M. Caspar, *ed.*, (München: Beck, 1940).

Kitaro, Nishida, (1966) *Intelligibility and the Philosophy of Nothingness*, R.S. Hinziger, *tr.*, (Honolulu: East-West Center).

Kojève, Alexander, (1964) "L'Origine Chrétienne de la Science Moderne," *Mélanges Alexander Koyré*, (Paris: Hermann).

Lawrence, Nathaniel, (1962) "Esthetic Creativity vs. Esthetic Appreciation," *Journal of the American Association of Museums*, (Oct.), v. 40, p. 20–24.

Lederberg, Joshua, (1966) "Experimental genetics and human evolution," *American Naturalist*, v. 100, p. 519–31.

Lenneberg, E.H., (1971) "Of Language Knowledge, Apes, and Brains," *Journal of Psycholinguistic Research*, v. 1, No. 1, p. 1–29.

MacKay, D.M., (1967) *Freedom of Action in a Mechanistic Universe*, (Cambridge: Cambridge University Press).

Marder, Leslie, (1971) *Time and the Space Traveller*, (Philadelphia: University of Pennsylvania Press).

McCulloch, W.C. and W. Pitts, (1943) "A Logical Calculus of the Ideas Immanent in Nervous Activity," *Bulletin of Mathematical Biophysics*, v. 5, p. 117–29.

Mill, J.S., (1843) *A System of Logic*, (London: Longmans, Green, and Co., Ltd., 1965).

Misner, C.W., K.P. Thorne and J.A. Wheeler, (1973) *Gravitation*, (San Francisco: Freeman).

Mumford, Lewis, (1963) *Technics and Civilization* (New York: Harcourt, Brace & World).

-, (1975) "Prologue to Our Time," *The New Yorker*, (March 10), p. 42-58.

Muses, C.E., (1965) "Aspects of some problems in biological and medical cybernetics," *Progress in Biocybernetics*, N. Wiener and J.B.S. Haldane, *eds.*, (New York: Elsevier), p. 243-48.

Needham, Joseph, (1944) *Time: the Refreshing River*, (London: Allen & Unwin).

-, (1954-77) *Science and Civilization in China*, (Cambridge: Cambridge University Press), 5 volumes.

-, (1961) "Human Law and the Laws of Nature," *Technology, Science, and Art: Common Ground*, (London: Hatfield College of Technology), p. 3-27.

-, (1966) "Time and Knowledge in China and the West," *The Voices of Time*, J.T. Fraser, *ed.*, (New York: Braziller), p. 92-135.

Oparin, A.I., (1924) "The Origin of Life," Ann Synge, *tr.*, reprinted in J.D. Bernal, *The Origin of Life*, (Cleveland: World, 1967), p. 199-234.

Prigogine, Ilya, (1973) "Time, Irreversibility and Structure," *The Physicist's Conception of Nature*, Jagdish Mehra, *ed.*, (Dordrecht and Boston: Reidel), p. 561-93.

Proust, Marcel, (1934) *Remembrance of Things Past*, (New York: Random House).

Rhind Papyrus. See: George Sarton (1952) *A History of Science, Ancient Science through the Golden Age of Greece*, (Cambridge: Harvard University Press).

Riemann, G.F.B., (1854) "On the Hypotheses which Lie at the Foundations of Geometry," *Habilitationschrift*, H.S. White, *tr.*, in *A Source Book in Mathematics*, D.E. Smith, *ed.*, (New York: McGraw-Hill, 1929), p. 411-425.

Roszak, Theodore, (1972) *Where the Wasteland Ends*, (New York: Doubleday).

Rothschild, F.S., (1970) "Eros and Thanatos in Human Evolution," *The Israel Annals of Psychiatry and Related Disciplines*, v. 8, p. 22-51.

Rowell, L.E., (1976) Personal communication.

Saunders, P.T. and M.W. Ho, (1976) "On the Increase in Complexity in Evolution," *Journal of Theoretical Biology*, v. 63, p. 375-84.

Schossberger, J.A., (1971) "The Individual as a Complex Open System," *Man in System*, M. Rubin, *ed.*, (New York: Gordon and Breach), p. 139-57.

Schuldt, A.C., (1976) "The Voices of Time in Music," *The American Scholar*, v. 45, p. 549-59.

Shannon, C.E. (1975) Quoted in "Computer Chess: Mind vs. Machine," *Science News*, v. 108, p. 345-50.

Sherington, C.S., (1947) *The Integrative Action of the Nervous System*, (New Haven: Yale University Press).

Sivin, Nathan, (1969) *Cosmos and Computation in Early Chinese Mathematical Astronomy*, (Leiden: Brill).

-, (1976) "Chinese Alchemy and the Manipulation of Time," *Isis*, v. 67, No. 239, p. 513-26.

Smith, Adam, (1776) *An Inquiry into the Nature and Causes of the Wealth of*

Nations. The quote is from Chapter II: "Of the Principles which give Occasion to the Division of Labor."

Solzhenitsyn, A.I., (1974) *Letter to the Soviet Leaders,* Hilary Sternberg, *tr.,* (New York: Index on Censorship).

Thass-Tienemann, Theodore, (1967) *The Subconscious Language,* (New York: The Washington Square Press).

Thom, René, (1975) *Structural Stability and Morphogenesis,* David Fowler, *tr.,* (New York: Benjamin).

Thorpe, W.H., (1961) *Bird Song,* (Cambridge: Cambridge University Press).

-, (1975) "Reductionism in Biology," *Studies in the Philosophy of Biology,* F.J. Ayala and Theodosius Dobzhansky, *eds.,* (Berkeley: University of California Press).

Turner, Frederick, (1971) *Shakespeare and the Nature of Time,* (Oxford: At the Clarendon Press).

-, (unpublished manuscript) *The Blaze at the Edge of the World: a Study in Poietic Time.*

Uexküll, Jakob von, (1921) *Streifzüge durch die Umwelten von Tieren und Menschen. Bedeutungslehre.* Georg Kriszat, *ill.,* (Frankfurt a.M.: Fischer, 1970). Both of these long essays are based on von Uexküll's *Umwelt und Innenwelt der Tiere. Streifzüge* ... is available in English as "A Stroll Through the Worlds of Animals and Men," C.H. Schiller, *ed.* and *tr.,* *Instinctive Behavior,* (New York: International Universities Press, 1957).

Voegelin, Eric, (1974) *The Ecumenic Age,* (Baton Rouge: Louisiana State University).

Vonnegut, Kurt, Jr., (1959) *The Sirens of Titan,* (New York: Dell).

von Neumann, John, (1951) "The General and Logical Theory of Automata," *Cerebral Mechanism in Behavior,* L.A. Jeffres, *ed.,* (New York: Hafner), p. 1-41.

Weber, Max, (1904) *The Protestant Ethic and the Spirit of Capitalism,* Talcott Parsons, *tr.,* (New York: Scribner, 1958).

Whorf, B.L., (1956) *Language, Thought, and Reality,* (Cambridge: M.I.T. Press).

Wigner, Eugene, (1967) *Symmetries and Reflections,* (Bloomington: Indiana University Press).

Wilson, E.O., (1975) *Sociobiology: the New Synthesis,* (Cambridge: Harvard University Press).

Winsor, Frederick, (1958) *The Space Child's Mother Goose,* (New York: Simon and Schuster).

Wright, Thomas of Durham (1750) *An Original Theory or New Hypothesis of the Universe,* (London: Chapelle).

Zeman, E.C., (1976) "Duffing's Equation in Brain Modelling," *Bulletin of the Institute of Mathematics and its Applications,* v. 12, p. 77-92. Contains references to Zeman's prior work.

Yourgrau, Wolfgang and C.J.G. Raw, (1968) "Variational Principles and Chemical Reactions," *Nuovo Cimento,* v. 5, Supp. 3 (1957), reprinted in W. Yourgrau and Stanley Mandelstam, *Variational Principles in Dynamics and Quantum Theory,* (Philadelphia: Saunders), p. 191-7.

Index

This index is for guidance only: it does not aspire for completeness. Slashes (/) are to be read as *and, or,* or *versus.*